NEW GUIDE TO THE PLANETS

PATRICK MOORE

NEW GUIDE TO THE PLANETS

SIDGWICK & JACKSON
LONDON

First published 1993 by Sidgwick & Jackson

a division of Pan Macmillan Publishers Limited
Cavaye Place London SW10 9PG
and Basingstoke

Associated companies throughout the world

ISBN 0 283 06145 6

9 8 7 6 5 4 3 2 1

A CIP catalogue record for this book is available from the British Library

Typeset by Spottiswoode Ballantyne Printers Ltd.
Printed by Mackays of Chatham.

CONTENTS

APPENDICES

FOREWORD

It is now forty years since I wrote the first edition of *Guide to the Planets*. Since then it has been revised and reprinted several times, but the last occasion was in the early 1980s, and it was clear that a complete re-write was called for. So much has happened during the past few years!

Many books about the planets have been produced recently, but I have not attempted here to go into great detail about the space-probe results; I am writing basically for the owner of a telescope who is anxious to do some real observing. I may well be accused of adopting an old-fashioned approach, and to this I plead guilty, but at least I am not covering the same ground as most other books now in circulation. I hope that some of my readers will feel spurred on to obtain a telescope and start looking at the planets for themselves. I know that they will not be disappointed at what they see.

Patrick Moore
Selsey, Sussex. 1993

CHAPTER ONE

The 'Wandering Stars'

We live in exciting times. The Earth is our home, but already some men have travelled far beyond it; a new era began in July 1969, when Neil Armstrong and Edwin Aldrin stepped out on to the surface of the Moon, and the gap between the two worlds had been well and truly bridged. True, they did not stay for long, and their journey was in the nature of a pioneer reconnaissance, but it was immensely significant. Already it seems a long time ago, and the Moon has not been visited since 1972, but there is every reason to believe that a fully-fledged Lunar Base will be set up in the foreseeable future.

Of course, this is only a beginning. The Moon, at its distance of only about a quarter of a million miles, is very much our nearest natural neighbour in space (excluding a few tiny asteroids, which can brush past us before receding once more), and it stays together with us as we travel round the Sun. It looks brilliant in our skies, and it needs an effort of the imagination to realize that it is a very junior member of the Sun's family or Solar System; it has no light of its own, and it shines only by reflecting the solar rays. Also, it is small. Represent the Earth by a tennis-ball, and the Moon will be no larger than a table-tennis ball.

The Earth itself is by no means important – except to ourselves. It is a normal-sized planet, unusual only because of its oxygen-rich atmosphere and its largely water-covered surface. We live here simply because conditions are right for us, and we are now sure that there is no other world in the Solar System upon which we could live in the open.

During the past few years, many books have been written about the planets – and there is plenty to say, bearing in mind the dramatic and often surprising information sent back by the various space-craft. My present aim is rather different, because what I want to do is to give an account of the planets from the point of view of the observer who is equipped with a small or moderate-sized telescope. So let me begin with a roll-call of the bodies which make up the Solar System. There are nine planets; their moons or satellites; comets, which are basically 'dirty ice-balls'; asteroids or minor planets; and an immense amount of general débris. And, of course, there is one star: the Sun.

Any glance at a plan of the Solar System shows that it is divided into two well-marked parts. First we have four relatively small, solid planets: Mercury, Venus, the Earth and Mars. Then comes a wide gap, in which move thousands of dwarf worlds known variously as asteroids, planetoids or minor planets. Next come the four giants: Jupiter, Saturn, Uranus and Neptune, plus a maverick dwarf, Pluto, which was discovered only in 1930 and which is probably not worthy to be ranked as a proper planet. Beyond Neptune and Pluto there is nothing but thinly-spread material until we reach the closest of the night-time stars.

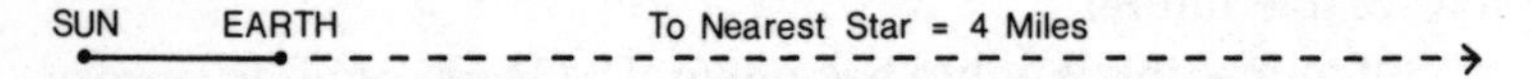

Distance of the nearest bright star beyond the Sun. If the Earth–Sun distance is represented by 1 inch, then α Centauri is over 4 miles away.

The Sun is 93,000,000 miles away,* which by astronomical standards is not very far. The other stars are so remote that their distances amount to millions of millions of miles. This means that for convenience we need a larger unit, and Nature provides us with one: the light-year. Light flashes along at 186,000 miles per second, so that in one year it covers a distance of rather less than 6 million million miles, which is what we term a light-year. Proxima Centauri, the first star we encounter beyond the Sun, is over four light-years away, corresponding to about 24 million million miles. It is interesting to work out a scale model, representing the Earth–Sun distance by one inch. How far away will Proxima then be? The answer: over four miles! The Earth–Sun distance, by the way, is known as the astronomical unit.

* In this book I propose to use Imperial units, so that everyone can understand them. If you prefer Metric, there is an easy way to turn miles into kilometres: simply multiply by 1.609.

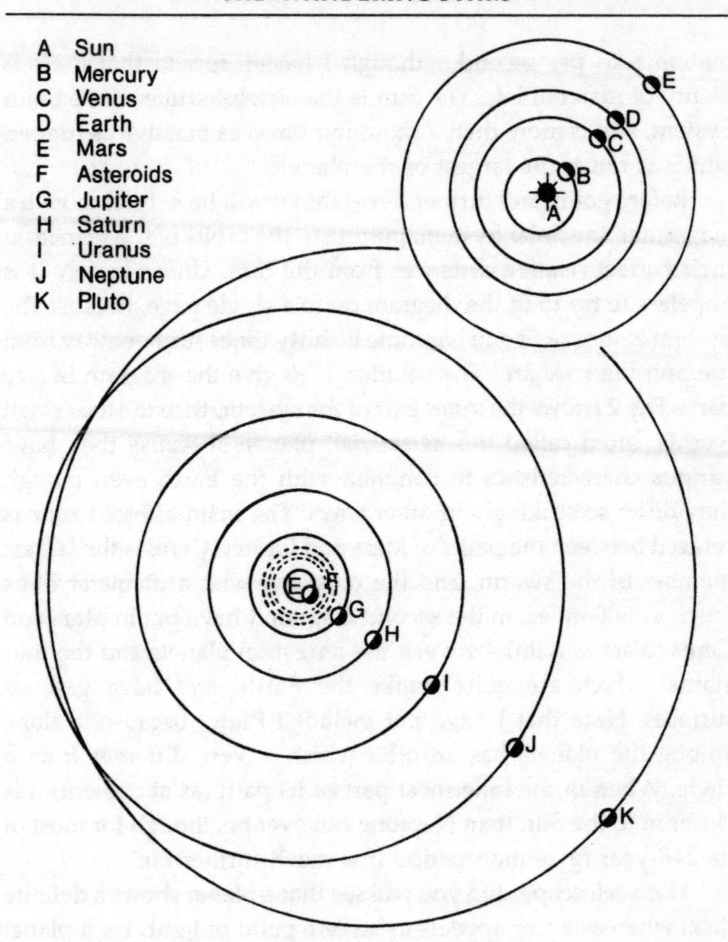

Plan of the Solar System, showing how it is divided into two parts separated by the asteroid belt.

With the naked eye the planets look like stars, but there is nothing genuinely starlike about them. The Sun is a vast globe of hot gas, with a surface temperature of almost 6000 degrees Centigrade (c. 11,000 degrees Fahrenheit) and a temperature at its core of 14 million degrees Centigrade (c. 25 million degrees Fahrenheit), possibly rather more; you could pack a million globes the volume of the Earth inside the Sun and still leave plenty of room to spare. It is creating energy by nuclear transformations taking place deep inside it, and it is losing mass at the rate of 4 million tons per second – though I hasten to add that there is plenty of material left. The Sun is the absolute ruler of the Solar System, and is more than a thousand times as massive as Jupiter, which is much the largest of the planets.

Before going any further, I feel that it will be helpful to give a diagram of the Solar System, putting in the orbits of the planets at their correct relative distances from the Sun. Unfortunately it is hopeless to try to fit the diagram on to a single page, because the System is so spread out; Neptune is thirty times further away from the Sun than we are. The solution is to give the diagram in two parts. Fig. 2 shows the inner part of the system, with the four small worlds, often called the 'terrestrial' planets, because they have various characteristics in common with the Earth even though they differ so strikingly in other ways. The main asteroid zone is centred between the paths of Mars and Jupiter; Ceres is the largest member of the swarm, and the only one with a diameter of as much as 500 miles. In the second diagram I have put in Mars and Ceres to act as a link between the terrestrial planets and the four giants, which are quite unlike the Earth, and have gaseous surfaces. Note that I have not included Pluto, because it alone among the planets has an orbit which is very different from a circle. When in the innermost part of its path (as at present) it is closer in to the Sun than Neptune can ever be, though for most of its 248-year revolution period it is much further out.

Use a telescope, and you will see that a planet shows a definite disk, whereas a star appears as a mere point of light. Each planet has its own special points of interest, and no two are alike. Of course, most of our present-day knowledge has been drawn from the space-craft, which have now by-passed all the planets except Pluto and have sent back information from close range; even so, we cannot claim to have solved all their mysteries – far from it. Surprisingly, perhaps, the amateur observer can still carry out useful research. Though I am bound to admit that the opportunities now are less than they used to be forty years ago, the systematic observer can make valuable contributions – and, occasionally, spectacular discoveries.

CHAPTER TWO

The Birth of the Planets

How was the world created? This is a question which mankind has been asking for many centuries, and there have been numerous theories, some of them plausible and others frankly wild. Nobody apart from the Biblical Fundamentalists has ever claimed to be able to give a full answer, but at least we are now modestly confident that we are on the right track. Certainly we have improved on the older ideas, but I particularly admire the views of the Iroquois Indians of North America, who explain everything very simply by saying that 'a heavenly woman was tossed out of Heaven, and fell upon a turtle, which grew into the Earth'. Nothing could be more straightforward!

Obviously we have to go back to the very beginning. It is now generally believed that the universe – space, time, matter, everything – was created in a 'Big Bang', between 15,000 million and 20,000 million years ago. If we are prepared to accept this, we can work out a fairly convincing evolutionary cycle, ending up with you and me. What we do not know, of course, is how or why the Big Bang happened. For the moment I propose to accept the majority view that it really did occur, though I agree that there are some much-respected scientists who remain sceptical. (The main problem is that it is difficult to find any sort of alternative.)

Galaxies formed; stars were produced in the galaxies. The age of our Sun is of the order of 5000 million years, so that it is not a 'first-generation' star; it has been formed from material sent out by older stars which have exploded in what we term supernova outbursts. Like most other bodies in the universe, the Sun is made up largely of hydrogen (over 70 per cent). Hydrogen is the lightest of all the elements; next in order, and also next in abundance, comes helium.

There can be no doubt at all that the Earth is younger than the Sun, and indeed we can fix its age fairly precisely at 4570 million years.* All the various lines of investigation lead to much the same result. Recorded history does not carry us far back into the past, but archaeological research can do much better, and studies of fossils – the remains of long-dead creatures – can tell us a great deal about what was going on hundreds of millions of years ago. Fossils, indeed, gave the first definite proof that the Earth is very old. Previously the Church had been regarded as the supreme authority on such matters, and in particular there was Archbishop Ussher of Armagh, who in 1654 stated that the world had been created at nine o'clock in the morning of 26 October, 4004 BC. (He reached this startling conclusion by adding up the ages of the patriarchs and making other equally irrelevant calculations. I have never been able to find out whether he made due allowance for Leap Years!)

However, fossils cannot take us right back to the beginning, because the Earth cannot have been born suddenly. The process must have been gradual, and life did not appear immediately, if only because the world was too hot. A more extended method of dating is known commonly as the radioactive clock, depending upon substances such as uranium.

Uranium, the heaviest of the ninety-two fundamental substances or elements found naturally on Earth, is not stable. It decays spontaneously, and finishes up as lead. It is in no hurry to disintegrate, and for one type of uranium, known scientifically as U^{238}, well over 4000 million years must pass before half of the original element has turned into lead. (With radium, this 'half-life' is only 1620 years.) Lead produced from uranium can be distinguished from ordinary lead, and so the quantity of uranium-lead associated with the remaining uranium tells us how long ago the decay started. This, in turn, gives a lower limit to the ages of the rocks containing the uranium.

Another radioactive element is rubidium, which ends up as strontium. Here the half-life is 46,000 million years, which is long even by cosmical standards. Note that the rate of decay of a

* Note that I am avoiding the use of the word 'billion'. The English billion is one million million, while the American billion is one thousand million. Most people today prefer the American version, but it can be misleading.

radioactive substance is constant; it is not affected by changes in temperature, pressure or anything else.

The results of radioactive dating seem to be quite definite. Some rocks are at least 4500 million years old, so that the world itself goes back further than this. Not so very many decades ago, such ideas would have been dismissed out of hand. For example Lord Kelvin, one of the great physicists of Victorian times, gave the age of the world as little over 10 million years; he arrived at this figure by calculating how long the Earth would take to cool down to its present temperature if it had originally been as hot as the Sun (surface temperature almost 6000 degrees Centigrade, 11,000 degrees Fahrenheit). I do not think there is any point in going into further detail here, because there really seems no doubt that our current figure of 4570 million years is very close to the truth. Man is a newcomer to the terrestrial scene, and historical periods seem very short on the timescale of the universe. If we represent the age of the Earth by one day, the Battle of Hastings was fought only one second ago.

The first really scientific theory was put forward in 1796 by Pierre Simon de Laplace, of France, who produced what is called the Nebular Hypothesis; in some ways it was not unlike earlier ideas proposed by Thomas Wright in England and Immanuel Kant in Germany, but was much more plausible. Laplace started with a picture of a vast gas-cloud, disk-shaped and in slow rotation. He then worked out an evolutionary sequence, ending up with a central sun attended by the planets and their satellites.

Laplace supposed that as the gas-cloud cooled down, radiating its heat away into space, it shrank; and the rate of spin increased until the centrifugal force at the edge became equal to the gravitational pull there. At this stage a ring of material broke away from the main mass, and gradually condensed into a planet. As the shrinking continued, a second broken-off ring produced a second planet, and the process was repeated over and over again, so that the end product was a central body (the Sun) attended by a retinue of circling worlds. The outermost planets were the oldest, and Mercury, the closest planet to the Sun, was the youngest member of the solar family.

At first sight the Nebular Hypothesis looks very neat, and it was generally accepted for many years, but then it was found that from a mathematical point of view there is a great deal wrong with it. The material shed during the shrinkage would not form separate rings – and even if it could, such a ring would not condense into a planet. Another trouble was that since the original gas-cloud was assumed to be flat, the orbits of the planets would lie in the plane of the Sun's equator, but they do not; the Earth's orbit is inclined to the equatorial plane of the Sun by a full seven degrees, and that of Mercury is tilted to ours at another seven degrees. But there is another difficulty which is even more serious.

If we look at one body revolving round another, and consider together its mass, its distance and its velocity, we arrive at what is known as angular momentum. It is a fundamental principle that angular momentum can be transferred, but never destroyed, so that on Laplace's theory all the angular momentum now possessed by the Sun and the planets must originally have been contained in the gas-cloud. At present, almost all the angular momentum in the Solar System resides in the four giant planets Jupiter, Saturn, Uranus and Neptune, but on the Nebular Hypothesis we would expect to find most of the angular momentum concentrated in the Sun itself. This would mean that the Sun would have to rotate fairly quickly. Actually it is a slow spinner, and takes over twenty-five Earth days to make a full turn. So although the Sun is 745 times as massive as all the planets put together, it accounts for only 0.5 per cent of the angular momentum of the whole Solar System.

By the end of the last century, mathematicians had launched so many savage attacks on the Nebular Hypothesis that it had to be cast on to the scientific scrap-heap. Next came a crop of theories involving both the Sun and a passing star. The first of these 'tidal theories' was proposed by Chamberlin and Moulton, in America, in 1900.

Space is very sparsely populated. The Sun, which is a normal star, has a diameter of over 860,000 miles, but if we represent it by a tennis-ball its nearest neighbours will be over a thousand miles away. This means that collisions or even 'close encounters' between two stars must be very rare indeed, at least in our part of the Galaxy. Two mosquitoes flying around inside the Royal Festival Hall would be unlikely to meet head-on, but the chances of them doing so are much greater than that of a stellar collision.

All the same, an encounter cannot be ruled out. On the

Chamberlin-Moulton theory, a wandering star approached the Sun so closely that tremendous gravitational interactions took place. Matter was pulled off the Sun, and remained as a cloud after the intruder had done its worst and receded into the distance. The material in the cloud began to condense into small bodies or 'planetesimals'; these in turn combined to form larger bodies, and the final result was the Solar System we know today.

Once a planetesimal had reached a diameter of a hundred miles or so, its own gravitational pull would become strong enough for it to collect extra material, and so it would grow relatively quickly. Unfortunately for the theory, it has been shown that under such conditions no planetesimal could ever become massive enough for the collecting process to begin. The high temperature of the material drawn away from the Sun would make the gases disperse long before they could start condensing.

Sir James Jeans, remembered today as much for his popular books and broadcasts as for his scientific researches, recognized this difficulty, and worked out a modified tidal theory according to which the planets condensed from a long, cigar-shaped filament pulled out of the Sun by the passing star. Certainly this would explain why the largest planets, Jupiter and Saturn, are more or less in the middle of the Solar System, since they would have been produced from the fattest part of the 'cigar', with the smallest planets at the tapering ends. The same sort of process could account for the satellite families of the giant planets, with the Sun cast in the disruptive rôle originally played by the wandering star.

Again the mathematicians swarmed to the attack. In an effort to improve matters, Sir Harold Jeffreys suggested that the intruding star actually struck the Sun a glancing blow. The idea was not entirely new – a somewhat garbled version of it had been published earlier by A. W. Bickerton, a rather eccentric New Zealander – and at first it looked promising. Then it too was critically examined, and fell by the wayside.

All in all, it now seems very doubtful whether any tidal or collision theory will work. We must start again.

The Sun, of course, is a solitary star, but stellar pairs are very common in the Galaxy. They are called binary systems, and are of various kinds. Sometimes the two components are perfect twins; sometimes one member of the pair is decidedly brighter and more

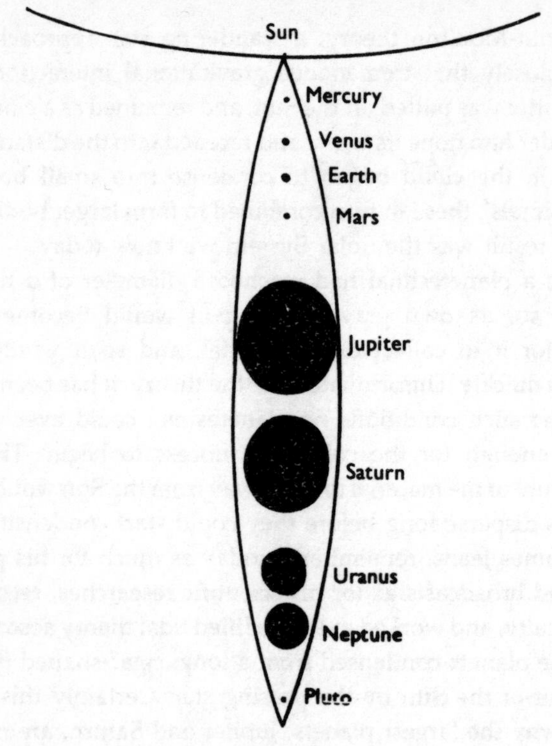

Jeans' tidal theory. The largest planets are in the middle of the 'cigar' pulled out of the Sun.

massive than the other; sometimes the components are strikingly unequal.

Probably the most famous of all binaries is Mizar, the second star in the 'tail' of the Great Bear (or, if you like, the handle of the Plough). Mizar is easily identifiable with the naked eye, because a much fainter star, Alcor, lies close beside it. When a telescope is used, Mizar itself is seen to be double. This is no mere line of sight effect; the components are genuinely associated, and are revolving round their common centre of gravity, much as the two bells of a dumb-bell will do when twisted by their joining bar. Sirius, the brightest star in the night sky, also has a binary companion, but small telescopes will not show it, since it has only 1/10,000 the luminosity of its brilliant primary.

Suppose that in the remote past the Sun was itself one

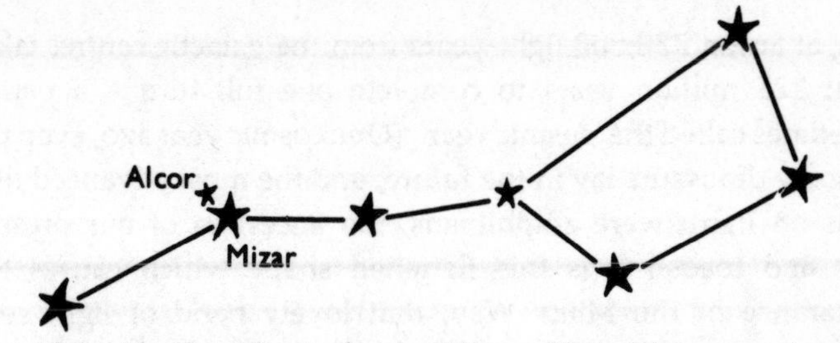

Ursa Major (the Great Bear), showing Mizar with its faint companion Alcor.

component of a binary system? H. N. Russell believed that it was the companion star which was struck by the intruder, causing enough débris to account for the planets; R. A. Lyttleton considered that a near approach by the wandering star would wrench the binary companion away from the Sun, planet-forming material being scattered in the process; Sir Fred Hoyle dispensed with the intruding star altogether, and suggested that the Sun's erstwhile companion exploded as a supernova, shedding much of its material before departing permanently. But none of these proposals met with much support. There was rather more enthusiasm for a theory due to Gerard Kuiper, who played such a major rôle in the early days of planetary exploration by space-craft. Kuiper regarded the Solar System as a degenerate binary, in which the second mass did not condense into a single body, but was spread out, so that the result was one star (the Sun) attended by a large number of condensations or protoplanets. The total mass of the protoplanets would amount to about one-tenth of the mass of the Sun, which is reasonable enough. Once formed, the protoplanets would contract to form the planets we know, drawing in other material from around and leaving the rest of the solar cloud to drift away in space. Unfortunately there is no real reason to believe that the Sun ever had a binary companion, degenerate or otherwise.

Today, the most favoured theories go back to some extent to the original Nebular Theory of Laplace. Of course they are very different in detail, but at least they do not involve any wandering star or hypothetical binary attendant.

Our Galaxy contains about 100,000 million stars. It is a flattened system, with a central bulge; it is in rotation, and the Sun, lying at around 30,000 light-years from the galactic centre, takes about 225 million years to complete one full turn – a period sometimes called the 'cosmic year'. (One cosmic year ago, even the fearsome dinosaurs lay in the future, and the most advanced life-forms on Earth were amphibians, the ancestors of our present frogs and toads!) It is this flattened shape which causes the appearance of the Milky Way, that lovely band of light seen stretching across the night sky. The stars in the Milky Way are not really crowded together, but in this direction we are looking along the main plane of the Galaxy, so that many stars are visible in almost the same line of sight.

The Galaxy has spiral arms, and if it could be seen from above or below it would look like a rather loose Catherine wheel, which is not at all surprising; spiral galaxies are very common in the universe. 'Pressure waves' appear to be the cause of the arms, and these sweep round the Galaxy. We live in a region affected by a pressure wave, in which gravitational shocks produce shrinking clouds of material which begin to fragment. Our Sun presumably began as one of these fragments – not initially as a star, if only because it was much too cool, but as a lens-shaped cloud or nebula about the mass of the present Sun, and with a diameter about as great as that of the orbit of the present-day Neptune. The temperature rose to a few tens of thousands of degrees, and the flattened lens began to separate into two parts: an inner 'proto-sun', and a much larger and more tenuous surround. Eventually the temperature at the core of the proto-sun reached about 10,000,000 degrees, and this was enough to trigger off nuclear reactions. The proto-sun started to shine; it had become a true star. The time taken for the cloud to evolve into a star was of the order of a hundred million years.

But what about the rest of the original cloud? It was from this material that the planets built up by the process known as accretion. Small lumps formed first, and these joined together to make 'protoplanets'; when they had become sufficiently massive they could draw in extra material from around. Near the Sun, the temperature was so high that volatiles such as ices were vaporized, which is why the inner planets are essentially rocky. Further out, where the temperatures were lower, the volatiles were not vaporized, and condensed on to the rocky cores to form the four giant planets.

Of course, this is a very sketchy and incomplete account of what happened (or rather, what we think happened), but it does account for many of the characteristics of the present Solar System. The fact that the solar nebula was lens-shaped explains why the planets move in very much the same plane; we can explain the reasons why the inner planets are so different from the giants, and why the distribution of angular momentum is what it is. We can also account for those strange, icy ghosts, the comets, which are presumably very ancient and were simply 'left over' from the main stages of formation.

There is one very important result of this change in ideas. Had a passing star been involved in the formation of the planets, Solar Systems would have been very uncommon; the stars are so widely spaced in most parts of the universe that stellar encounters are vanishingly rare. But if the planets were produced from a solar nebula, there is every reason to assume that such systems are common. After all, the Sun is a very ordinary star, and there is nothing in any way exceptional about it. And if there are 'other Earths', why not other civilizations?

One day we may find out. Meanwhile, we have at least a feeling of confidence that we are on the right track.

CHAPTER THREE

The Movements of the Planets

At first glance, a planet looks very much like a star. True, each planet has its own special characteristics; Venus and Jupiter stand out at once because they are so bright, and Mars because it is so red – but Saturn at all times, and Mars when furthest from the Earth, do look remarkably star-like when seen with the naked eye. It is not even true to say, as many books do, that a star twinkles while a planet does not. Twinkling is due entirely to the Earth's unsteady atmosphere, which 'shakes the light around', so to speak, with the result that twinkling is much more evident with a star which is low over the horizon. A planet shows up as a tiny disk rather than a point source, and so the twinkling is less, but it is not absent.

Of course, the reason why ancient astronomers could single out the planets was because of their motion against the starry background. Originally it was believed that everything in the sky moved round the Earth, and this scheme of things was brought to its highest degree of perfection by Ptolemy of Alexandria, who lived between about AD 120 and 180. His system is always known as the Ptolemaic, though to be accurate Ptolemy himself did not invent it.

On the Ptolemaic theory, the Earth was motionless, and lay in the centre of the universe. Round it, in perfect circles, moved the Moon, Mercury, Venus, the Sun, Mars, Jupiter and Saturn. Beyond Saturn came the sphere of the fixed stars. All celestial orbits were circular, because the circle is the 'perfect' form, and nothing short of absolute perfection can be allowed in the heavens. Of course, the sky turned round the Earth once in twenty-four hours.

The main problem was that the movements of the planets in the sky did not fit in with any simple, uncomplicated theory. Mars, Jupiter and Saturn sometimes perform slow 'loops', taking many

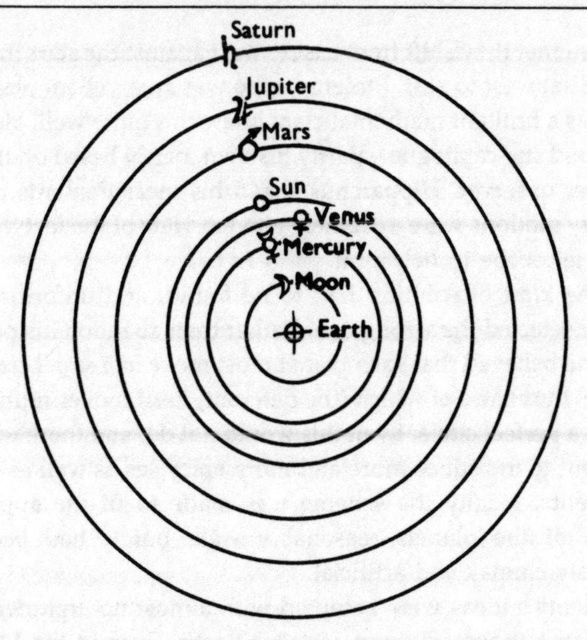

The Ptolemaic System; the planets move in perfectly circular orbits round a stationary Earth.

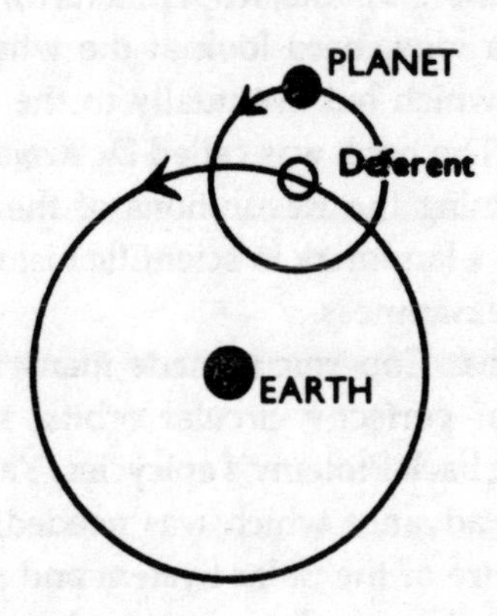

Epicycles and deferents. On the Ptolemaic theory, a planet moves in a small circle or epicycle, the centre of which (the deferent) itself moves round the Earth in a perfect circle.

weeks, so that they shift from east to west against the stars instead of the usual west to east. Ptolemy, who was an excellent observer as well as a brilliant mathematician, knew this quite well. He also had a good star catalogue – partly his own, partly based on that of an earlier observer, Hipparchus – and his measurements of the planetary motions were extremely good in view of the fact that he had no telescope to help him.

Some kind of solution had to be found, and unfortunately Ptolemy selected the wrong one. Rather than abandon his perfect circles, he believed that each planet must move in a small circle or epicycle, the centre of which (the deferent) itself moves round the Earth in a perfect circle. Even this would not do, and there was no choice but to introduce more and more epicycles as well as other refinements. Finally the scheme was made to fit the apparent motions of the planets reasonably well, but it had become hopelessly clumsy and artificial.

Ptolemy's ideas were accepted with almost no argument for more than a thousand years after his death. Even in the Middle Ages, any proposal to dethrone the Earth from its proud central position was regarded as heretical. Then, in the mid-sixteenth century, a Polish canon, Mikołaj Kopernik (always known to us as Copernicus) took a long, hard look at the whole situation, and published a book which led eventually to the overthrow of the Ptolemaic system. The book was called *De Revolutionibus Orbium Cœlestium* (Concerning the Revolutions of the Celestial Bodies) and it proved to be a landmark in scientific history. It also led to a great deal of unpleasantness.

Let us admit that Copernicus made many mistakes. He still kept to the idea of perfectly circular orbits, and he was even reduced to bringing back Ptolemy's epicycles. Yet he made the one great fundamental advance which was needed; he removed the Earth from the centre of the Solar System and put the Sun there instead. From being the most important body in the universe, our world was relegated to the status of an ordinary planet.

This sort of heresy did not appeal to the Christian Church, which was never slow to take drastic action against anybody who disagreed with it. Copernicus was a priest, and was under no delusions – so he avoided trouble by the simple means of withholding publication of his book until the very end of his life.

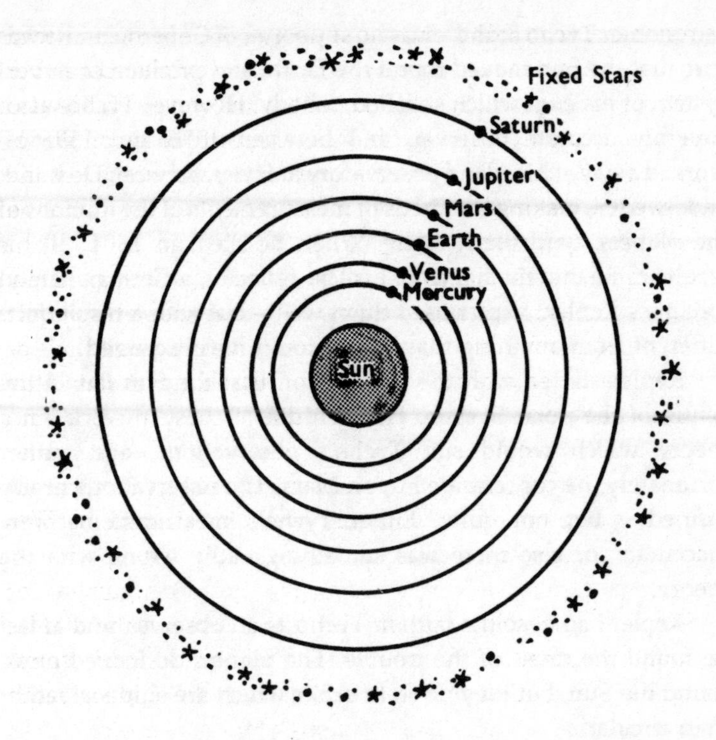

The Copernican System. The Sun lies in the centre of the system, with the planets moving round it in perfectly circular orbits.

Some of his later supporters were not so prudent. In 1600 Giordano Bruno was burned at the stake in Rome, one of his crimes being that he persisted in teaching the Copernican theory rather than the Ptolemaic. And – to run slightly ahead of the story – there is the case of Galileo, the first great telescopic observer, who was brought to trial in the 1630s and forced by the Inquisition to 'curse, abjure and detest' the false theory that the Earth moves round the Sun! Incredibly, it was only in 1992 that the Catholic Church finally admitted that Galileo had been right all along, and even then it took a special Vatican commission, set up by Pope John Paul II in the 1970s, thirteen years to come to a definite decision.

Oddly enough, the man who provided the full observational data which destroyed the Ptolemaic theory – the Danish astronomer Tycho Brahe – was no supporter of Copernicus; he was sure that the Sun moved round the Earth, and produced a hybrid system of his own which satisfied nobody. However Tycho was a superbly accurate observer, and between 1576 and 1596 he worked away at his island observatory at Hven, between Denmark and Sweden, making hundreds of measurements of the motions of the planets, particularly Mars. When he died, in 1601, all his results came into the hands of his last assistant, a German named Johannes Kepler. Kepler used them well – but with a result quite different from anything that Tycho could have expected.

Kepler started with the assumption that the Sun lies in the centre of the Solar System. He then did his best to work out a theory which would suit Tycho's observations, and, rather fortunately, he concentrated upon Mars. The observations nearly fitfitted – but not quite. Either Tycho's measurements were inaccurate, or else there was something badly wrong with the theory.

Kepler had absolute faith in Tycho as an observer, and at last he found the cause of the trouble. The planets do indeed move round the Sun, but they do so in orbits which are elliptical rather than circular.

One way to draw an ellipse is to stick two pins into a board, an inch or two apart, and join them with a thread, leaving a certain amount of slack. Now draw the thread tight with a pencil, and trace a curve. The result will be an ellipse, the two pins marking the foci. If the pins are wider apart, with the same length of thread, the ellipse will be longer and narrower. The distance between the foci is a measure of the eccentricity of the ellipse.

In the case of the Earth, the Sun occupies one focus of the ellipse, while the other focus is empty. (I realize that this is another over-simplification, because we have to reckon not only with the Sun–Earth pair but also with the Moon and all the other planets, but the general scheme is clear enough.) The Earth's path is nearly circular, and completely unlike the eccentric ellipse shown in the left-hand diagram, which is more like the orbit of a comet. The Earth's distance from the Sun never ranges more than $1\frac{1}{2}$ million miles from the average value of 93,000,000 miles. In the right-hand diagram I have drawn the orbit to the correct eccentricity – and you will agree that it looks very circular! Yet it was the slight

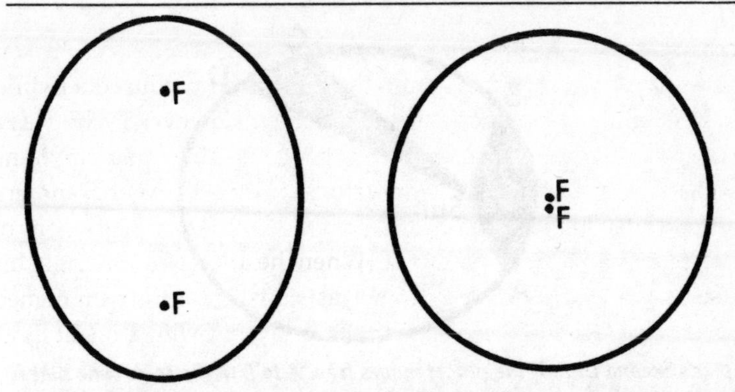

Ellipses. (Left) *An eccentric ellipse, similar to the orbit of a short-period comet.* (Right) *An ellipse which is almost circular, such as the orbit of the Earth. F mark the foci of the ellipse; with a body moving round the Sun, the Sun occupies one focus of the ellipse, while the other is empty.*

departure from the circular form which led Kepler on to his great discovery. It was lucky that he had such complete faith in the accuracy of Tycho's observations of the movements of Mars.

Mars has an orbit which is rather more eccentric than ours, so that its distance from the Sun ranges from approximately 128,500,000 miles at its closest approach or *perihelion* out to approximately 154,500,000 miles at its furthest or *aphelion*. If the path of Mars had been as nearly circular as that of (say) Venus, it would have taken Kepler much longer to find the answer to his problems.

Once he had taken the decisive step, Kepler was able to draw up three fundamental Laws of Planetary Motion, the first two of which were published in 1609 and the third in 1618. The first law states that a planet moves round the Sun in an ellipse, with the Sun at one focus. The second states that the *radius vector* – that is to say, a line joining the centre of the planet to the centre of the Sun – sweeps out equal areas in equal times, while the third gives a relationship between the planet's orbital or sidereal period and its distance from the Sun.

The second law is worth a little further explanation. In the diagram, the shaded area is equal to the dotted area, assuming that the planet takes the same time to move from A to B as it does from C to D. In other words, a planet moves quickest when at peri-

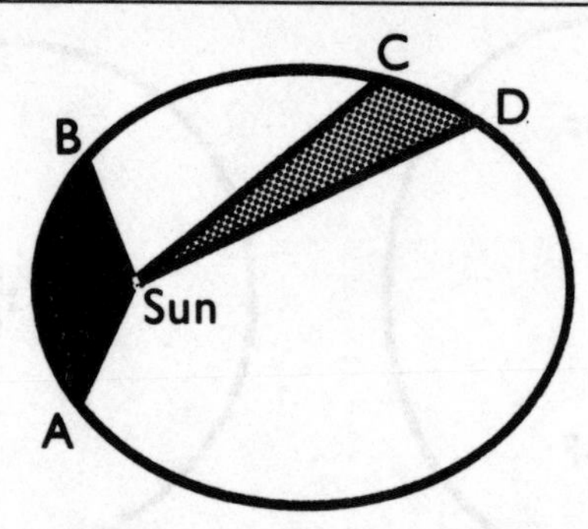

Kepler's Second Law. If the planet moves from A to B in the same time that it takes to move from C to D, then the blacked area A-Sun-B must be equal to the shaded area C-Sun-D.

helion. It also follows that the nearer planets are the faster-moving; Mercury races round the Sun at a mean velocity of 30 miles per second, the Earth at 18.5 miles per second* and Jupiter 8 miles per second, while Neptune has a mean orbital velocity of only just over 3 miles per second.

Two of the planets, Mercury and Venus, are closer to the Sun than we are, so that they have their own way of behaving; they are often called the 'inferior planets'. To begin with, we can never see them against a really dark background except when they are low down in the sky. They always lie roughly in the same part of the sky as the Sun, and are at their brightest when in the west after sunset or in the east before dawn. Moreover, they show phases similar to those of the Moon. The diagram explains just what happens.

Let us take Mercury first. When at position 1, it has its dark side turned toward us, and is 'new', so that it cannot be seen at all; this is known as *inferior conjunction*. As it moves along in its orbit, a little of the daylight side begins to be turned toward us, so that Mercury comes into view. It shows up as a crescent; then, at position 2, it is a half-disk (*dichotomy*) and then becomes *gibbous*, or three-quarter phase, before becoming full at position 3. At full, or *superior conjunction*, it is almost behind the Sun, so that it is difficult to see even with a telescope. After superior conjunction,

* 18.5 miles per second = 66,000 mph.

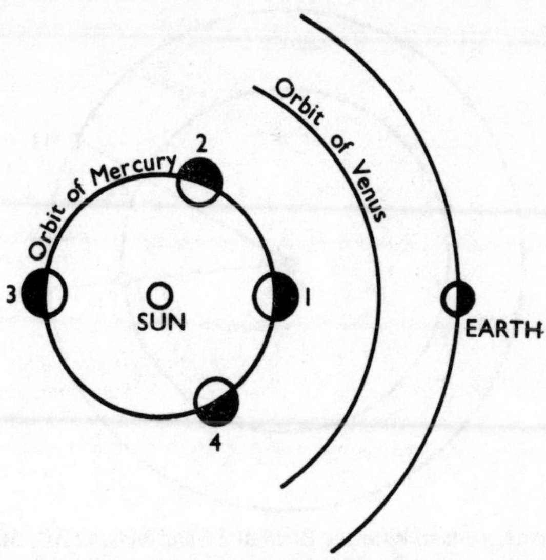

Phases of Mercury. (1) New. (2) Dichotomy (half-phase). (3) Full. (4) Dichotomy. For the sake of clarity, I have not taken the Earth's movement round the Sun into account in this diagram.

the phases are repeated in the reverse order: gibbous, half at position 4, crescent, and then back to new. When an evening object, Mercury is on the wane; when a morning object, it is waxing. Because of its quick motion, it passes through all its phases several times a year. In 1993, for instance, it was new on 9 March, 15 July and again on 6 November. (Of course, the Earth's own movement round the Sun has to be considered, but I have not taken this into account in the diagram.)

The angular separation between Mercury and the Sun is never as much as 30 degrees, and it never rises much before the Sun or sets long after it. Also it is a small world, not a great deal larger than the Moon, and as the phase increases the apparent diameter shrinks, because Mercury is drawing away from the Earth. City-dwellers will probably never see it with the naked eye, though people living in the country may see it shining fairly brightly when it is best placed.

At some inferior conjunctions Mercury passes directly between the Sun and the Earth, so that for a few hours it may be seen as a black spot in transit across the brilliant solar disk. Transits

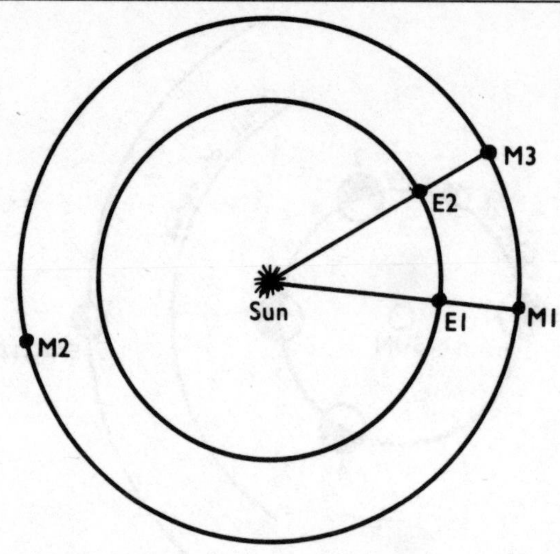

Movements of Mars. With the Earth at E1 and Mars at M1, Mars is at opposition. A year later the Earth has come back to E1, but Mars has only reached M2. The Earth has to catch Mars up before there is another opposition, with the Earth at E2 and Mars at M3.

fell on 10 November 1973, 13 November 1986 and 6 November 1993; the next will be on 15 November 1999. Transits can only happen in May or November, and do not occur at every inferior conjunction because Mercury's orbit is tilted with respect to ours, allowing Mercury to pass unseen either above or below the Sun in the sky.

Venus behaves in the same way as Mercury, but is much more conspicuous, partly because it is further from the Sun and partly because it is much larger and more reflective than Mercury. At its best it can cast strong shadows, and there are times when it is still visible five and a half hours after sunset or before sunrise. Transits are very rare, and the next will not fall until the year 2004.

The remaining or *superior* planets lie beyond the orbit of the Earth in the Solar System, and are much more convenient to observe. Mars is shown in the diagram given here, but again I have not taken the orbital eccentricity into account.

Obviously, no planet outside the Earth's orbit can ever pass through inferior conjunction, since it can never go between the Sun and the Earth. At the corresponding position in its orbit Mars, at M1, is certainly lined up with the Sun and the Earth, but this time the Earth is in the mid position, with Mars opposite to the Sun in the sky and best placed for observation. This is termed *opposition*. With Mars, the average interval between successive oppositions (the *synodic period*) is 780 days. To a Martian observer, the Earth would then be at inferior conjunction.

Like Mercury and Venus, Mars and the outer planets can reach superior conjunction. In this position Mars is virtually behind the Sun as seen from Earth, so that it is above the horizon only in broad daylight, and for a few weeks it is to all intents and purposes out of view.

The reason why Mars comes to opposition only once in 780 days or so is because both it and the Earth are moving. The Earth takes just over 365 days to make one circuit of the Sun, and if we start with the Earth at E1 and Mars at M1, the Earth will have returned to E1 in 365 days. Mars, however, is moving more slowly in a larger orbit, so that it has a longer 'year' – 687 Earth-days – and it will not have arrived back at M1. It will only have travelled as far as M2. The Earth has to catch it up, which will happen with the Earth at E2 and Mars at M3. This means that oppositions of Mars do not fall every year; for example there is no opposition in 1994.

Conditions are not quite the same for the more distant planets. They move so slowly compared with the Earth that they are much easier to catch up, so that they have shorter synodic periods, and come to opposition at intervals of little over a year. For example, Jupiter's synodic period is 399 days, so that the opposition of March 1993 is followed by another in April 1994. Pluto's synodic period is 366.7 days, so that oppositions occur only two or three days later each year.

We can now account for the occasional backwards or *retrograde* movements which so perplexed Ptolemy. In the next diagram, the orbits of the Earth and Mars are shown, with the apparent positions of Mars in the sky. Between positions 3 and 5 the Earth is catching Mars up and passing it, so that for a while Mars seems to 'loop the loop'.

The orbits of the planets (discounting Pluto) lie very much in the same plane, so that if we draw a plan of the Solar System on a flat piece of paper we are not far wrong. This means that the planets can never wander near the poles of the sky; they keep

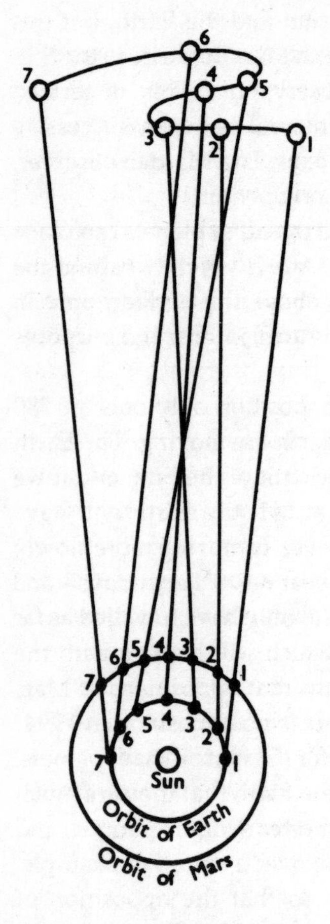

Retrograde movement of Mars. As the Earth catches up Mars and passes it, the movement will seem to be retrograde, so that between 3 and 5 Mars will appear to move backwards in the sky – east to west, against the stars, instead of west to east.

relatively close to the *ecliptic,* which is the plane of the Earth's orbit, and may also be defined as the apparent yearly path of the Sun against the stars. The ecliptic passes through the twelve constellations of the Zodiac; Aries (the Ram), Taurus (the Bull), Gemini (the Twins), Cancer (the Crab), Leo (the Lion), Virgo (the Virgin), Libra (the Scales or Balance), Scorpius (the Scorpion), Sagittarius (the Archer), Capricornus (the Sea-goat), Aquarius (the Water-bearer) and Pisces (the Fishes). A thirteenth constellation, Ophiuchus (the Serpent-bearer) does cross the Zodiac for some distance between Scorpius and Sagittarius. Astrologers, who believe (or say they believe) that the positions of the planets in the Zodiac at the time of a person's birth have an effect on his or her character and destiny, have a positive hatred of Ophiuchus; they pretend that it doesn't exist. However, it is fair to say that astrology proves only one scientific fact: 'There's one born every minute!'

I must stress that this brief account of the motions of the planets can give little idea of the enormous problems which face the mathematical astronomer – even in the age of computers. Each planet pulls upon the others, producing disturbances or *perturbations;* we have to take the satellites of the Earth and the giants into account, and there are any number of other complications, such as the 'relativity' effect predicted by Albert Einstein – which has turned out to be particularly important in the case of Mercury.

Astronomical handbooks publish tables of the planets, but remember that these are only for the mean orbits, because a planet does not follow exactly the same track in every journey round the Sun. An orbit depends mainly on the velocity of the planet, and this velocity is not constant. Taking every predictable perturbation into account it is possible to compute an *osculating* orbit, but this is only for one particular moment. For example, consider the year 1961, when we were just starting to consider sending rockets out to the planets. Neptune is a case in point. In early 1961 it was moving according to an orbit whose eccentricity was slightly less than the mean value. Drawn on a diagram which would fit on to this page, the difference between the two orbits would be concealed by the breadth of an inked line, but the actual difference between the calculated 'mean' position of Neptune at that time (computed according to the 'mean' orbit) and the real position (computed according to the osculating orbit) amounted to no less than 9,000,000 miles.

The complications seem endless, and uncertainties and errors are bound to remain even today. Yet we can be sure that our results are much more accurate than those of a few decades ago. Otherwise, we could not hope to launch a successful rocket probe out to a planet far away in the Solar System.

CHAPTER FOUR

Rockets to the Planets

There are two ways of exploring the planets. One – the age-old method – is by telescope; by observing from the surface of the Earth we can learn a great deal. The other is by means of spacecraft. The first step was taken in 1959, when the Russians launched three probes towards the Moon, and followed it up two years later with the first attempt to reach Venus, even though this particular rocket was a failure. Since then we have been able to obtain close-range views of all the planets apart from Pluto, and controlled landings have been made on Mars and Venus.

In a book dealing with the planets themselves, I do not think that I should say much about rocket theory, because it is so much part of our everyday life (which is more than could be said when the first edition of the book appeared, way back in the early 1950s.)* Therefore I will be brief, but I must say something, because even now there are some widespread misconceptions about how a rocket actually works.

To begin with, it does not 'push against' anything except itself. A November the Fifth rocket consists of a hollow tube filled with gunpowder, plus a stick to give stability in flight. When the powder burns, it gives off hot gas; this gas rushes out of the exhaust, and propels the rocket skyward. The rocket flies because of what Sir Isaac Newton so rightly called the principle of reaction: every action has an equal and opposite reaction. So long as the gas continues to stream out, the rocket will go on moving.

The essential point here is that the Earth's atmosphere is not concerned. Indeed, it is actually a nuisance, because it sets up

*I still have a copy of a review of the first edition of this book, written by an eminent astronomer who shall remain nameless. 'Even though Moore's idea of sending rocket vehicles out to Mars and Venus is so obviously naïve . . .'

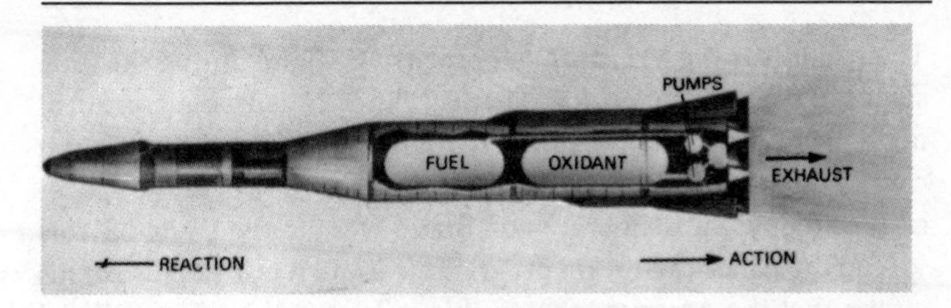

Principle of a rocket.

friction and has to be pushed out of the way. Rockets are at their best in 'empty' space, where there is virtually no resistance at all. This is why rockets, and rockets alone, can be used for flight beyond the Earth.

The first man to realize this clearly was a Russian, who rejoiced in the name of Konstantin Eduardovich Tsiolkovskii. He published his first papers almost a hundred years ago, but they caused no excitement at all, partly because they appeared only in obscure Russian periodicals and partly because the whole idea of space-travel was officially regarded as crazy. Tsiolkovskii also realized that solid fuels, such as gunpowder, are of no use for sending probes directly from the Earth to the Moon or anywhere else, because they are too weak and too uncontrollable. Instead he planned to use liquid propellants, and to replace the crude charge of gunpowder with a proper rocket motor. Another of his ideas was to mount rockets one on top of the other, so that the upper 'step' would be given a running start into space. All this was perfectly sound, but Tsiolkovskii himself never fired a rocket in his life; he was not a practical experimenter, and activities of such a kind would probably have been unpopular in Tsarist Russia. It was only in 1926 that the first liquid propellant rocket was sent up, not by Tsiolkovskii but by an entirely independent American scientist, Robert Hutchings Goddard. It was a modest affair, travelling for a short distance upward at a mere 60 mph, but it showed that the principle was valid.

Further experiments were made in the 1930s, mainly in Germany by a team led by Wernher von Braun. They looked very promising – so promising in fact, that the Nazi Government stepped in, and put the rocket men to work on military weapons.

The result was the V.2, used to bombard England during the final stages of the war. In 1944 and early 1945 V.2s caused a great deal of damage, and many people have rueful memories of them, but it cannot be denied that they were the direct ancestors of the spacecraft of today. Subsequently von Braun went to the United States to lead the American rocket programme – and it was because of his efforts that the Americans were able to launch their first artificial satellite, Explorer 1, in 1958.

But this is running ahead of the story. Early rockets, such as the V.2s, could not break free from the pull of the Earth because they could not go fast enough. Here let me dispose of another popular misconception. No rocket can leave the Earth's gravitational field, which is theoretically infinite. In theory it is possible to go from the Earth to the Moon in a spacecraft moving at no more than a few miles per hour, but you would have to go on using fuel all the way, which would be hopelessly uneconomical (quite apart from the fact that the journey would take a very long time indeed). The only practical method is to give the vehicle an initial acceleration which will break it free from the Earth's pull without further expenditure of fuel. There is some analogy, though not an accurate one, with a cyclist who pedals so hard that he can then coast up a hill without using any further energy.

The critical velocity is 7 miles per second, or around 25,000 mph. If the rocket can accelerate to this speed, it will not fall back to the ground, and can 'coast' toward the Moon. This is why 7 miles per second is termed the Earth's escape velocity.

This was no new idea even in Tsiolkovskii's time. One of those who had grasped it was Jules Verne, the great French storyteller, who wrote a novel about a lunar voyage as long ago as 1865. Old-fashioned though it may be, it is a splendid story and well worth reading, but Verne made the fundamental mistake of launching his travellers in a projectile fired from the muzzle of a huge gun. This involved a sudden departure at escape velocity, and there are several reasons why this is absolutely out of the question. Any vehicle moving at such a speed through the dense lower layers of the Earth's air would be destroyed by friction, and in any case the shock of departure would have very unfortunate consequences for anyone who happened to be inside the projectile!

With a rocket probe the acceleration can be gradual, so that the vehicle reaches full escape velocity only when it has passed beyond the thick part of the Earth's atmosphere. Tremendous power is needed, and this is why we have to adopt the 'step' principle first outlined by Tsiolkovskii. Immediately after blast-off, the large lower motor* does all the work; when it has used up all the propellant it can carry, the whole of the bottom stage of the launcher breaks away and falls back to the ground. The motors in the second stage are fired, and when they too are exhausted the motors in the third stage take over, so that only the top part of the entire vehicle goes the whole way. With the Apollo Moon programme, the pre-launch vehicle stands over 360 feet in height, equal to St Paul's; yet only the 22-foot cone carrying the astronauts came back intact at the end of the mission.

The Moon, of course, is a special case, because it is our companion and always stays with us as we travel round the Sun. A trip there and back takes less than a fortnight. When we turn to the planets, we are faced with much greater problems, because the distances are so much greater and also because the planets do not obligingly stay close to us.

Since Venus was the first planet to be chosen as a target, I think it is appropriate to begin there; the diagrams will help. In the left-hand diagram Venus is at V1 and the Earth is at E1; the minimum distance between the two may be less than 25,000,000 miles, which is roughly a hundred times as far as the Moon. Why not wait for a suitable moment, with Venus at inferior conjunction, and fire a probe straight across the gap?

Alas, this cannot be done. Without continuous application of power no vehicle can move in a path of this kind, and the amount of propellant needed would be prohibitive. Moreover, both Venus and the Earth are moving, and no space-ship can manœuvre in the same way as an aircraft.

The procedure actually followed is rather different. Basically, what is done is to take the probe up in a step-vehicle, and then slow it down relative to the Earth. Remember that once a vehicle is in space, and unpowered, it moves in exactly the same way as a natural body would do, and it obeys Kepler's Laws. When the probe has been 'slowed', and the motors switched off, it will not

* Or, rather, motors. The design of a spacecraft is amazingly complicated, and the account I have given here has been watered down to the point of dehydration.

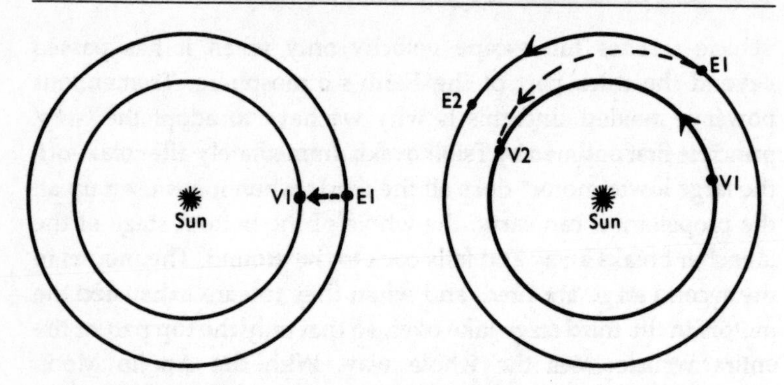

Orbit of Venus probe. (Left) *Impracticable path, from the Earth (E1) straight to Venus (V1).* (Right) *Transfer orbit. The spacecraft is launched when the Earth is at E1 and Venus at V1. The probe swings inwards, and meets Venus at V2. Meanwhile the Earth has moved on to E2.*

stay with the Earth; it will begin to swing inward towards the Sun, picking up speed again as it does so. Eventually it will reach the orbit of Venus. If all the calculations have been correct, it will rendezvous with Venus as shown in the right-hand diagram. If Venus were not there to meet it, the probe would continue to swing inwards until it had picked up enough velocity to start moving outward once more, and it would go on moving round the Sun in an elliptical orbit for an indefinite period.

To land on Venus, the motors in the probe must be used at the critical moment. Alternatively, the probe can be put into a closed path round the planet, as has been done with the latest spacecraft, Magellan, about which I will have more to say later.

Most of the journey from the orbit of the Earth to that of Venus has to be done in unpowered free fall – or coasting, if you like – and is bound to take time. Mariner 2, the first successful Venus explorer, was launched on 27 August 1962 and did not reach its closest point to the planet until 14 December. Magellan was launched on 5 May 1989 and did not arrive until 10 August 1990. Eventually, no doubt, we will be able to speed things up by using different kinds of propellants, but for the moment we have to rely upon liquids which are not nearly as powerful as we would like.

With Mars, the problems are of the same basic type, but this time the probe has to be speeded up relative to the Earth, so that it swings out towards the Martian orbit, losing speed steadily. If it does not land, or is not put into a closed path around Mars, it will continue indefinitely in an elliptical orbit round the Sun. Here too the first success was American; Mariner 4 was sent up on 28 November 1964, and reached the neighbourhood of Mars over seven months later, on 15 July 1965.

With the more remote planets the times of travel become ominously long. The first Jupiter probe, Pioneer 10, was launched on 2 March 1972 and did not arrive until 3 December of the following year. But with the outer giants there is one way of cutting the time of travel: the method known officially as 'gravity assist', but which I always think of as interplanetary snooker. It is possible to use the gravitational pull of one planet to send a probe on to its next target.

Consider Voyager 1, which was launched towards Jupiter in September 1977. It passed the Giant Planet on 5 March 1979 at a minimum distance of just over 200,000 miles, and used Jupiter's

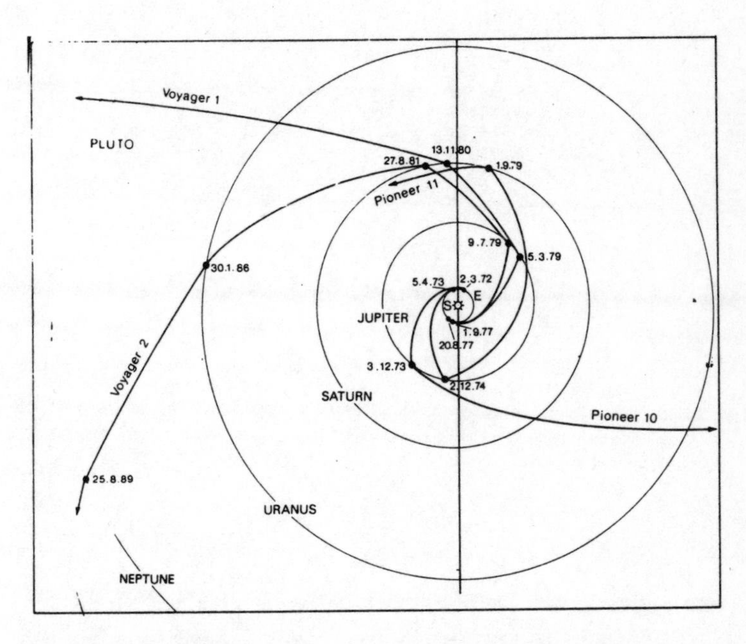

Trajectories of the Voyagers. (S = Sun.) Voyager 1 encountered Jupiter and Saturn; Voyager 2 added Uranus and Neptune as well.

immensely powerful gravitational pull to throw it into a path which took it on to a rendezvous with Saturn on 12 November 1980. After that, it began a never-ending journey which is taking it right out of the Solar System. At the moment we are still in touch with it, and we know exactly where it is, but in a few years we will certainly lose track of it.

Voyager 2, its twin, had an even more amazing career. It took in all the four giants: Jupiter in 1977, Saturn in 1981, Uranus in 1986 and finally Neptune in 1989. This was possible because, by sheer good luck, the four giant worlds were spread out in a long, gentle curve – a situation which will not recur for a very long time.

These various spacecraft have revolutionized all our ideas about the planets. Moreover, it has all happened very quickly. During the past thirty years we have been able to find out more about our neighbour worlds than we had been able to do all through human history.

CHAPTER FIVE

Mercury

Of all the naked-eye planets, Mercury is much the least obtrusive. There must be many people who have never seen it, because it has to be looked for in just the right place at just the right time. People who live in modern cities or industrial areas have no hope at all. Incidentally, there is a legend that the great astronomer Copernicus never glimpsed it because of mists rising from the river Vistula near his home, but I am quite sure that this is apocryphal – if only because I have seen Mercury very easily from that area, and there are many more artificial lights around now than there were in the sixteenth century.

Mercury can actually be more brilliant than any of the stars – even Sirius. The trouble is that we can never see it against a dark sky. It always stays inconveniently close to the Sun, so that it is a naked-eye object only for brief periods when it is very low in the west after sunset or very low in the east before sunrise. From my home in Selsey, on the Sussex coast, I have found that the average number of nights per year when Mercury is visible with the naked eye is between fifteen and twenty, but Sussex has clearer skies than most other parts of Britain, and I also have a sea horizon. One word of warning: do not sweep for the planet with binoculars or a telescope unless the Sun is completely below the horizon. Bring the Sun into the field of view by mistake, even for an instant, and you will ruin your eyesight.

From more southerly countries Mercury is easier to see, and the ancients knew it well. Originally they thought that 'evening Mercury' and 'morning Mercury' must be two different bodies, but before long they realized that the two are identical. Because the planet is so quick-moving it was named in honour of Hermes, the messenger of the gods; the Latin equivalent of Hermes is Mercury, and the 'geography' of Mercury is still officially termed 'hermography'.

Until recent times we knew little about its surface features. It is small, with a diameter of only 3032 miles; it is on average 36,000,000 miles from the Sun, so that it never comes much within 50 million miles of us. Virtually all our information about the surface comes from one spacecraft, Mariner 10, which made three surveys in 1973 and 1974. But before discussing Mercury itself, can we be sure that there is no major planet closer in to the Sun?

Little more than a century ago, it was believed that such a planet really existed. It was even given a name: Vulcan, after the blacksmith of the gods. The story of this so-called 'discovery' provides an excellent example of how even the most brilliant scientist can be misled.

The Director of the Paris Observatory during the 1870s was Urbain Jean Joseph Le Verrier. He was probably the most eminent astronomer of the day, and he had achieved lasting fame by his success in tracking down the planet Neptune in 1846. In 1860, Le Verrier studied the movements of Mercury, and found that something was wrong. The 'messenger of the gods' was not quite where he should have been, and Le Verrier decided that Mercury was being pulled out of position by the gravitational force of some unknown planet.

Then there came a message from a French country doctor, Lescarbault, who claimed that he had watched an intra-Mercurian planet passing in transit across the face of the Sun. Remember that at some inferior conjunctions both Mercury and Venus show up in transit, and presumably an inner planet would do the same – which, incidentally, would be almost the only hope of observing it at all. Le Verrier made haste to visit Lescarbault. The interview must have been a curious one. Le Verrier had the reputation of being one of the rudest men in France, while Lescarbault was hardly a professional scientist; he doubled as the village carpenter, and wrote down his observations with chalk on planks of wood, planing them off when he had no further use for them, while his timekeeper was an ancient watch which had lost its second hand. Yet in spite of all this, Le Verrier accepted the results as confirming the existence of the planet. He named it Vulcan, and worked out its distance from the Sun as 13,000,000 miles, with a sidereal period of between 19 and 20 days and a diameter of about 1000 miles. He also calculated the times of future transits.

Yet Vulcan has never been seen since, and it is now certain that what Lescarbault recorded was not a planet. It may have been a sunspot, but it is worth noting that E. Liais, from Brazil, had been observing the Sun at the exact time of Vulcan's supposed transit, and had seen nothing at all.

Interest was rekindled briefly in 1878, when there was a total eclipse of the Sun. At a total eclipse the Moon passes directly in front of the Sun and blots out the bright disk, so that for a few minutes stars can be seen with the naked eye in the middle of the day. Two American observers, Watson and Swift, carried out a careful search, and claimed that they had recorded various unidentifiable objects. Alas, Watson's and Swift's observations agreed neither with the predicted Vulcan nor with each other, and presumably what they saw were nothing more than ordinary stars. Since then the irregularities in the movements of Mercury have been cleared up without the need to bring in a new planet; Einstein's theory of relativity has provided a full explanation. There is no Vulcan. Let us return, then, to Mercury.

There is no problem in working out its movements and its mass. The revolution period round the Sun is 88 days – one Mercurian 'year'; the diameter, as we have noted, is 3032 miles, as against 2158 miles for our Moon, and the mass is 0.06 per cent that of the Earth. The escape velocity is 2.6 miles per second, and the surface gravity 0.38 that of the Earth. If your Earth weight is 12

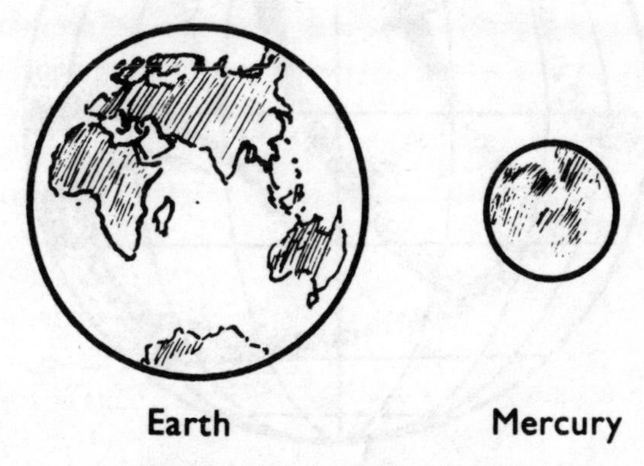

Mercury and the Earth compared.

stones, you will find that on Mercury you will be reduced to an ethereal 3 stones.

What we did not know, before the epic flight of Mariner 10, was anything really definite about the surface features, though lunar-type mountains and craters were fairly confidently expected. Quite apart from its smallness and its distance, Mercury is an awkward object to observe from Earth. When closest to us it is at inferior conjunction, with its dark side turned toward us; as the phase grows, the apparent diameter shrinks. When full, the planet is on the far side of the Sun, and cannot be seen. Telescopic observers have always had to battle against heavy odds.

The first systematic observations were made in the late eighteenth century by William Herschel, but with little effect. At about the same time J. H. Schröter, in Germany, used his telescopes to claim that high mountains rose from the Mercurian surface – but with due respect to Schröter, who was totally honest, it is difficult to take these results seriously. So we come to Giovanni Virginio Schiaparelli, an energetic Italian, who began his series of observations in 1881 and went on until 1889. He was based at Milan, and used refractors of aperture 8.6 inches and 19.8 inches.

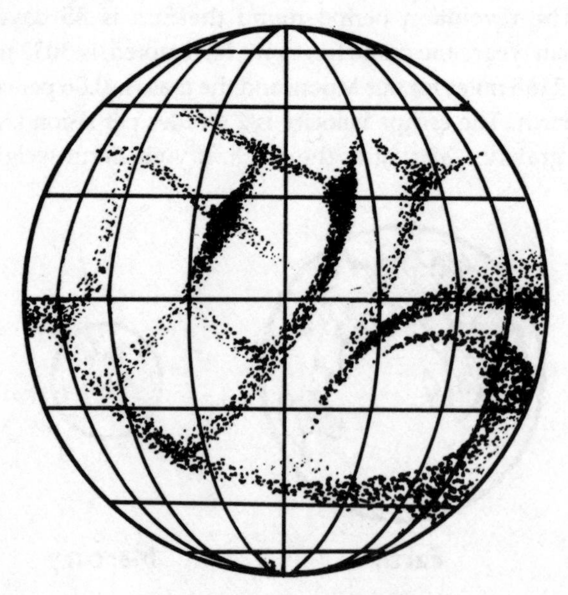

Schiaparelli's map of Mercury.

Schiaparelli knew, of course, that when Mercury is visible with the naked eye it is bound to be low over the horizon, so that seeing conditions are bound to be poor. The only alternative is to observe during broad daylight, when Mercury is high up. This was Schiaparelli's method, and eventually he produced a map of the planet whch showed what he believed to be definite bright and dark areas. He came to the conclusion that the revolution period was the same as the rotation period – 88 Earth days – in which case Mercury would present the same face to the Sun all the time. This was mathematically sound, and, after all, the Moon behaves in precisely this way with respect to the Earth – a state of affairs known technically as captured or synchronous rotation. There is no mystery about it in the case of the Moon; tidal effects over the ages are responsible, and Schiaparelli believed that the Sun's pull had similarly 'braked' the spin of Mercury.

His results were duly acclaimed. Next in the line of great planetary observers came Eugenios Antoniadi, Greek by birth but French by adoption, who had the advantage of being able to use one of the world's largest and best refractors: the 33-inch at Meudon, near Paris. I can vouch for the quality of this superb instrument, because I used it a great deal in the far-off days when I was busily mapping the Moon before the Apollo astronauts went there. (I also used it to look at Mercury, though I cannot pretend to have seen anything definite on the tiny disk.)

Antoniadi was in full agreement with Schiaparelli. In his view the rotation was indeed synchronous, but it was not true to say that exactly half the planet would be in permanent sunlight and the other half in permanent darkness, because Mercury's orbit is not circular; it is markedly eccentric, and the distance from the Sun ranges between 28.5 million miles at perihelion out to 43.4 million miles at aphelion. This means that the oribital velocity varies – but the rotational speed does not. Therefore, the position in orbit and the amount of spin will become regularly out of step, and Mercury will 'rock' very slowly to and fro, just as the Moon does to us. (These effects are called librations; I will have more to say about them in Chapter 8.) On Mercury there would be a region of permanent day, a region of everlasting night, and between them a relatively narrow 'twilight zone' over which the Sun would bob up and down across the horizon. Later science fiction writers made

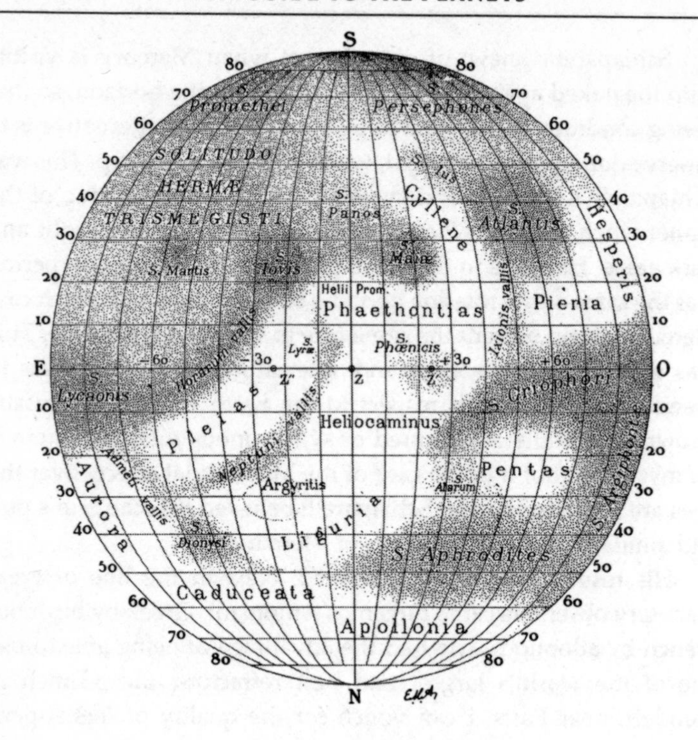

Antoniadi's map of Mercury.

great use of the twilight zone; it would be the only part of Mercury where temperature conditions would be at all tolerable.

In 1934 Antoniadi produced a book about Mercury, in which he included a map of the surface.* He gave names to the dark features: Solitudo Hermæ Trismegisti (the Wilderness of Hermes the Thrice Greatest), Solitudo Criophori and so on. Solitudo Hermæ Trismegisti was described as 'a vast shading, which I discovered on 17 August 1927 and which I have since seen almost continuously, even with very bad images. This desert-like tract extends over the south-east quadrant of Mercury and, with its extensions, is comparable in size with Australia . . . To me, the colour always seems grey, like all the other dark patches on Mercury.'

* Surprisingly, the book was available only in French until 1974, when I produced an English translation (Keith Reid Ltd, Devon, 1974). By that time it was, of course, of historical value only. The quotes given in this chapter are from my translation.

More controversial was Antoniadi's claim that Mercury possessed an appreciable atmosphere. From the outset, this did not seem very likely. The power of a body to hold down an atmosphere depends upon two factors: the temperature, and the escape velocity. The Earth, with its equable temperature and its escape velocity of 7 miles per second, can hold down a dense atmosphere (fortunately for us!). The Moon, with an escape velocity of 1½ miles per second, is quite unable to do so. Mercury, with an escape velocity of 2.6 miles per second, should be just about on the critical limit, but it is very hot – and the higher the temperature, the greater the speeds of atmospheric molecules, and the greater the probability of escape.

Yet Antoniadi maintained that there were frequent obscurations over Mercury, and that the clouds were 'more frequent and obliterating' than those of Mars, sometimes persisting for several days and covering dark areas such as Solitudo Criophori. There was never any doubt about Antoniadi's skill as an observer, and for many years nobody else was able to carry out so careful a study with so large a telescope, but uneasy doubts about his 'clouds' remained.

What, then, about life? Antoniadi was prepared to admit that 'it is possible, albeit not probable, that very rudimentary life-forms, such as microbes, cannot be absolutely ruled out from the polar regions of Mercury', and it is fascinating to look back at his description of the Mercurian scene as given in his book:

> 'An observer on the planet – assuming that he could exist there – would first find that his weight had fallen to one-third of its Earth value; he would walk slowly, and jump high; stones thrown would fall silently at great distances across the endless deserts which cover all parts of the planet. It would be difficult for him to hear his own voice, since sounds would be so much weakened by the rarefaction of the atmosphere. Shadows would appear, on average, 2½ times more intense than those on Earth. The dusk and dawn so peculiar to Mercury would seem excessively pale, but of very long duration, because of the slowness with which the Sun would rise and set.
>
> 'A more-than-glacial breeze would be extremely punishing during the night, while a wind incomparably more scorching

than the desert simoon would probably give rise to a spectacle of fuming dunes, which during the day would raise eddies of greyish dust which would cover the sky, and conceal the Sun with sinister, all-obscuring clouds. Moreover, there would be literally no humidity, evaporation or condensation; no cirrus, stratus, cumulus, nimbus, waterspouts or mists; neither rain, ice, snow, dew or white frost. Next, the absence of water-droplets and ice-needles in the air would mean that the observer would see no rainbows, haloes, parhelia or mock-suns; there would not even be coronæ; and in spite of the very great heat of the shifting Sun, it is doubtful whether mirage effects could transform the desert plains into optical lakes.

'If volcanic eruptions still occur, they might help in producing beautiful rosy twilight effects, similar to those which lit up our skies after the Krakatoa disaster of 1883. The intensity of possibly electrical phenomena is difficult to estimate in the atmosphere of a world so near the Sun, and which has neither moisture nor water. However, even if there are no magnetic phenomena, and if the composition of the atmosphere is as we have assumed, the monotony of the eternal Mercurian night will often be broken by magnificent displays of aurora polaris, which will be particularly brilliant near the time of sunspot maximum.

'The constellations will, of course, present the same aspect from Mercury as from the Earth; but they will show up more brilliantly, and the Milky Way will appear so low in altitude that it will seem to be nearly on top of the observer. Star-twinkling will be unknown. From the equator of Mercury, a star will set not after 12 hours, but after 44 Earth-days. The celestial sphere will appear to turn from east to west with majestic slowness.

'The Zodiacal Light, in which Mercury is always plunged, will seem to be more luminous but more diffuse than it does to us; and since it will fill most of the sky, it will form a very broad band, somewhat ill-defined; it will be 360 degrees in extent, visible even in the daylight hemisphere, with its axis coinciding approximately with the plane of the solar equator. A pale, luminous condensation at 180 degrees from the Sun will correspond to the feeble counterglow of our winter nights.

'Seen from Mercury, all planets are naturally superior, coming to opposition with the Sun. Venus will be the most brilliant star in the sky, much more brilliant than it appears from Earth – a veritable celestial diamond, casting shadows with diffraction fringes. Its disk at opposition will attain a diameter of 70 seconds of arc when Mercury is near aphelion. Our Earth will come next, a magnificent star of the first magnitude, whose maximum diameter will attain 33 seconds of arc, and which, with its bright surface, should be comparable with our own view of Venus, casting comparable shadows. Keen eyesight would also show our Moon, slowly oscillating from one side of the Earth to the other. Mars, as seen from Mercury, would never fall to the second magnitude. Jupiter and Saturn would be of the first magnitude for most of the time, but sometimes fainter; and Uranus would be slightly less brilliant than as seen from Earth.

'Mercury would be an ideal planet from which to admire and study comets. Indeed, these "hairy stars" would appear, near their perihelion, with a grandeur which would make our comets of 1811, 1843, 1858 and 1882 seem very feeble.

'Meteor showers ought to be rich from Mercury; but as the planet's orbital velocity is 48 kilometres per second, as against the 30 kilometres per second of the Earth, and meteors moving at the distance of Mercury will also be travelling faster, it follows that shooting-stars will, in general, describe very rapid trajectories in the rarefied Mercurian atmosphere. If the planet were habitable, meteorite falls would be dangerous, since the thin atmosphere would give only inadequate protection against a celestial bombardment.

'Another consequence of the rarefaction of the atmosphere is that it has relatively little diffusive power, so that on Mercury stars would be visible in full daylight. Thus Venus, the nearest planet, which can approach Mercury to within 39,000,000 kilometres, would remain visible whenever it were above the Mercurian horizon and were not hidden by clouds. It would be seen even when close to the Sun. During the day the Earth would be easily visible, and so, though rather less in evidence, would Jupiter, Mars and Saturn.

'Finally, an observer on Mercury who looked at the Sun with the naked eye, near the time of perihelion – with the Sun low in the sky, and seen through thick clouds of desert dust – would be able to make out not only the large spots, but also any others

> which subtended an angle of over 12 seconds of arc, provided that these small spots were almost circular; and it is interesting to add that the enormous solar disk would itself appear circular, without sensible flattening, because the atmospheric refraction at the surface of Mercury would be so slight.'

Poetic, plausible – and, alas, completely wrong. We now know that Mercury is not in the least like this, and the discoveries made during the last few decades show that it is totally hostile.

The first revelation came in 1962, as a result of work by W. E. Howard and his colleagues from Michigan. They measured the long-wavelength radiation coming from Mercury – the infra-red – and found that the dark side was not nearly as cold as it would have been if it never received any sunlight. Confirmation came from radar studies carried out by Rolf Dyce and Gordon Pettengill, using the powerful 'dish' set in a natural bowl at Arecibo in Puerto Rico. Basically, radar works by sending out a pulse, bouncing it off a solid body or something which acts in an equivalent way, and recording the echo. The results tell us a great deal about the object off which the pulse has bounced. There is some analogy with a tennis-ball thrown against a wall and caught on the rebound; you can learn at least something about the wall. I do not pretend that the analogy is a good one, but at least it may have some relevance.

Mercury is a small, elusive target, but by the mid-1960s it was well within radar range. When pulses are reflected from a body which is spinning round, the echo is modified, and the rate of rotation can be found. The Arecibo team found that the rotation period is not 88 days after all; it is 58.7 Earth days. There is no permanently sunlit zone, no region of everlasting night, and – rather sadly – no twilight zone.

It is reasonable to assume that in the early period of the Solar System Mercury span more quickly than it does now, and that it was slowed down by the Sun's gravitational pull. But there is another curious relationship which may or may not be due to sheer coincidence. I do not want to delve into mathematics, even very simple ones, so I will try to sum the situation up as concisely as I can:

1. The synodic period of Mercury – that is to say, the time which elapses between successive appearances at the same phase

– is, on average, 116 Earth days. If Mercury is 'new' on a particular date, it will again be new 116 days later.

2. The rotation period (58.7 Earth days) is equal to two-thirds of the revolution period (88 Earth days).

3. It follows that to an observer at a fixed position on Mercury, the interval between one sunrise and the next will be 176 Earth days, or two Mercurian years.

4. This interval, 176 Earth days, is approximately equal to $1\frac{1}{2}$ synodic periods.

5. From this, it will be found that after every three synodic periods, the same face of Mercury will be seen at the same phase.

6. Now for the coincidence – if coincidence it be! Three synodic periods of Mercury add up to approximately one Earth year. Consequently, the most favourable times for looking at Mercury recur every three synodic periods. Glance back at Point 5. You will realize that every time Mercury is best placed for observation, we see the same hemisphere, with the same markings in the same positions on the disk.

Moreover, the Mercurian calendar is decidedly weird. As we have seen, the orbit is markedly eccentric, so that the orbital speed ranges between 36 miles per second at perihelion down to 24 miles per second at aphelion. The axial inclination is negligible – only about 2 degrees – so that Mercury spins almost 'upright' with respect to its orbit, as against a planet such as the Earth, where the tilt is $23\frac{1}{2}$ degrees to the perpendicular. When Mercury is near perihelion, the orbital angular velocity exceeds the constant spin angular velocity, so that an observer on Mercury would see the Sun slowly retrograde or move backwards through rather less than its own apparent diameter for eight Earth days around each perihelion passage. The Sun would then almost hover over what may be called a 'hot pole'. There are two hot poles, one or the other of which will always receive the full blast of solar radiation when Mercury reaches perihelion; the intensity will be $2\frac{1}{2}$ times that absorbed by regions of the surface 90 degrees of longitude away. Bear in mind, too, that from Earth the Sun has an apparent diameter of only half a degree; from Mercury, it will range between 1.1 and 1.6 degrees.

Let us consider two observers, both of whom are placed on Mercury's equator, but who are 90 degrees in longitude away from

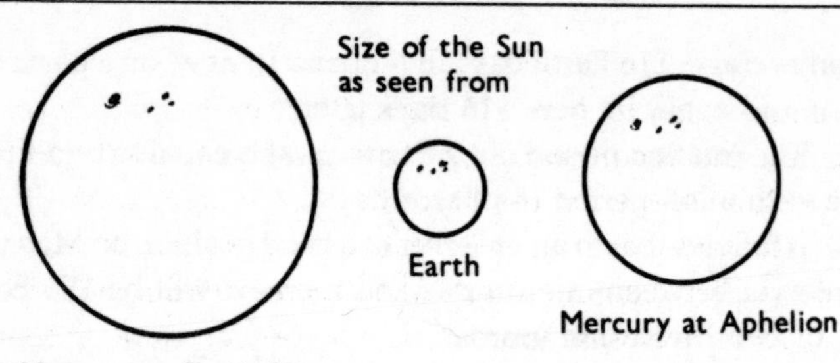

Relative size of the Sun as seen from Earth compared with the view from Mercury at perihelion and aphelion.

each other. Observer A is at a hot pole, so that the Sun is at his zenith, or overhead point, at perihelion. This means that the Sun will rise when Mercury is near aphelion, and the solar disk will be at its smallest. As the Sun nears the zenith, it will slow down and grow in size. It will pass the zenith, and then stop and move backwards for eight Earth days before resuming its original direction of motion. As it drops toward the horizon it will shrink, finally setting 88 Earth days after having risen.

Observer B, 90 degrees away, will see the Sun at its largest near the time of rising, which is also Mercury's perihelion. Sunrise itself will be curiously erratic, because the Sun will come into view and then sink again until it has almost vanished. Then it will climb into the sky, shrinking as it nears the zenith. There will be no 'hovering' as it passes overhead, but sunset will be protracted; Mercury is back at perihelion, so that the Sun will disappear, rise again briefly as though bidding adieu, and then finally depart, not to rise again for another 88 Earth days.

Another odd fact is that the stars will move across the sky at roughly three times the average rate of the Sun. One hates to think what a Mercurian would make of all this!

By the late 1960s, then, we had found out that Mercury was not quite the sort of world we had expected it to be, but we were no nearer to finding out anything about its surface features, and various maps drawn over the years were no improvement upon Antoniadi's. The next step came in 1973, with the launch of the first – and so far, the only – space-probe to visit Mercury: Mariner 10.

It was sent up on 3 November, and obtained some good

pictures of the Moon before swinging inwards to an encounter with Venus on the following 5 February. It by-passed Venus at 3600 miles, and used the gravity-assist method to put it into an orbit which made it fly past Mercury on 29 March; the first pictures, showing craters and mountains, were actually received five days earlier, when Mariner was still three million miles from Mercury. After its closest approach the space-craft went on round the Sun, and came back to a second encounter with Mercury on 21 September. The third rendezvous came on 16 March 1975, but by that time the on-board equipment was starting to fail, and on 24 March contact was finally lost. No doubt Mariner 10 is still orbiting the Sun, and still making regular passes of Mercury, but whether it will ever be re-contacted is more than doubtful; to all intents and purposes it is dead, but certainly it had done everything that its planners had expected of it.

It could not, of course, map the entire surface. The same region of the planet was in sunlight during each pass (the area covered by the Solitudo Hermæ Trismegisti in Antoniadi's map; the Solitudo Criophori area was out of view). Even now we have charted less than half the surface, though there is no reason to believe that the remainder is basically different.

No serious attempt was made to retain Antoniadi's nomenclature; through no fault of his, his chart was simply not accurate enough. The naming committee of the International Astronomical Union worked out an entirely new system. Craters have been named after personalities, ranging from musicians (Beethoven, Chopin, Wagner) to writers (Mark Twain, Dickens, Shakespeare), artists (Gainsborough, Renoir, Van Gogh) – in fact men and women from almost every walk in life apart from politicians, who have been very sensibly excluded. Ridges have been named after famous ships of exploration, and valleys after radar installations.

At first sight the surface of Mercury, as shown by Mariner 10, is very like that of the Moon, but there are important differences in detail. Mercury does not have the broad lava-plains which we still miscall the lunar seas, but there are many intercrater plains, which are fairly level, rolling areas with a rough texture due to vast numbers of craterlets. There are also 'lobate scarps', long, sinuous cliffs running for hundreds of miles, characterized by a rounded appearance unlike anything on the Moon. The most prominent feature of all is the Caloris Basin, 800 miles across; it has been so named because it lies at one of Mercury's hot poles, with the Sun overhead at perihelion. It is bounded by a ring of smooth mountain blocks rising to at least a mile above the surrounding surface in places. Unfortunately, only part of the Basin was sunlit during the Mariner passes.

Some of the craters are very large; one, Beethoven, has a diameter of almost 400 miles. Small craters, less than about a dozen miles across, are bowl-shaped, while others have central peaks and terraced walls. There are also ray-craters, similar to those of the Moon. In fact, one of these – now named Kuiper, in honour of the space-probe pioneer Gerard Kuiper – was the first feature to be identified as Mariner 10 closed in towards the planet. Crater Kuiper is almost forty miles wide.

There can be virtually no activity on Mercury now. However the craters were formed – by meteoritic bombardment, as most astronomers believe, or by internal action, as is claimed by a minority – there must have been extensive vulcanism long ago; according to the most reliable estimates the age of the Caloris Basin is around 4000 million years, and the main active period ended not long afterwards. Mercury today is a barren, changeless place.

As expected, the atmosphere proved to be negligible, with a ground density of no more than one thousand-millionth of a millibar; the main constituent seems to be helium, presumably captured from the solar wind. More surprisingly, there is a weak but detectable magnetic field, which is of the same basic type as ours, even though it is only one-hundredth as powerful. The magnetic axis is inclined to the rotational axis by about 14 degrees. The presence of a magnetic field is due to the fact that Mercury has a relatively large, heavy, iron-rich core, which may be greater than the whole globe of the Moon. The core may or may not be molten – we have no firm evidence either way – and the crustal layer is 'loose' down to a depth of a few feet, making up what is commonly called a 'regolith'.

One very startling announcement was made in 1991, not from the Mariner results but from work carried out by radar astronomers in New Mexico. Here we find the VLA or Very Large Array, made up of 27 antennae spread over a wide area. Very high resolution can be obtained, particularly when the VLA is used in conjunction with other installations, and Mercury was selected as a target. It was found that strong radar echoes coming from near the planet's north pole could be attributed to nothing more nor less than ice!

An ice-cap would hardly be expected on a world such as Mercury, but it is true that there are craters near the poles which have permanently shadowed floors and must be intensely cold. However, I admit to being profoundly sceptical. It does not seem likely that there has ever been water on Mercury, and without water there can be no ice.

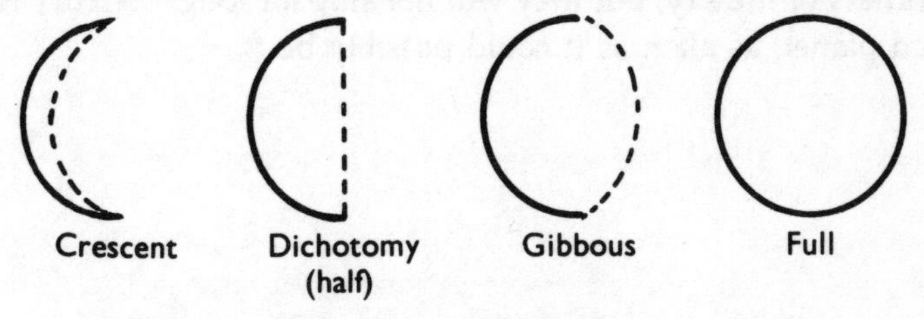

Limb and terminator. The limb is shown as a solid line; the terminator is dashed.

So far as the amateur observer is concerned, we have to agree that Mercury is far from promising. The main interest is in following the changing phase, and noting the position of the terminator, or boundary between the sunlit and night sides of the planet. The horns of the crescent are termed cusps.

I have made many attempts to see features on Mercury, and I have had the advantage of being able to use some of the world's most powerful telescopes, but I have never had much luck, and I doubt if anyone could do much better – it would be difficult to match Antoniadi. Obviously it is best to follow Schiaparelli's method of observing in broad daylight, but this means having accurate setting circles; casual sweeping around with binoculars or a telescope is emphatically not to be recommended, because of the ever-present danger of looking at the Sun by mistake.

Transits across the face of the Sun occur now and then, always in May or November. For example, transits during the last decade of the present century fall on 6 November 1993 and 15 November 1999. Strange phenomena have been noted during past transits – bright spots on the planet's disk, for example – but it seems that these are due to defects either in the instruments or in the observer's eye. During transit, Mercury is too small to be seen with the naked eye, so that observations have to be made by using the telescope as a projector. It is worth comparing Mercury with any sunspots which happen to be on view; Mercury will look inky black, whereas a sunspot does not.

All in all, Mercury is one of the most hostile of all the members of the Sun's family. One day it is conceivable that men will go there, and see the huge, glaring Sun shining down from the darkness of the sky, but they will not stay for long. Mercury is a dead planet, as alien as it could possibly be.

CHAPTER SIX

Venus

Venus, the second planet in order of distance from the Sun, is very different from Mercury. In fact, the only point that the two have in common is that both are very hot. With its diameter of 7523 miles, Venus is almost the same size as the Earth; it is an excellent reflector of sunlight, and is the closest natural body in the sky apart from the Moon and an occasional passing asteroid or comet. It moves round the Sun in an almost circular orbit at a mean distance of 67,200,000 miles, so that at its nearest to us it is only about a hundred times as far away as the Moon. At its most brilliant, it can cast strong shadows.

Like Mercury, it stays in the same region of the sky as the Sun, but the angular distance between it and the Sun can be as much as 47 degrees, so that Venus can remain visible for as much as five and a half hours after sunset or rise five and a half hours before the Sun does so. It can then be seen against a dark background, and is truly magnificent. It is not surprising that the ancients named it after the Goddess of Beauty.

Unfortunately, it is a telescopic disappointment, because the true surface is permanently hidden by a deep, cloudy atmosphere. Hard, sharp markings, such as those of Mars, are conspicuous only by their absence. Moreover, when Venus is closest to us – at inferior conjunction – it has its dark side turned Earthward, and we cannot see it all except during the very rare transits. Full phase occurs with Venus on the far side of the Sun, so that again it is to all intents and purposes out of view. Maximum brilliance occurs when about 30 per cent of the sunlit side faces us. During the crescent stage, keen-eyed people can see the phase under ideal conditions, and good binoculars show it easily.

The phases themselves have been known for a long time; they

The changing size of Venus as seen from Earth. It is closest to us when new, furthest away when full.

were recorded by Galileo in 1610. They can be predicted, since the movements of Venus are known very accurately – and yet theory and observation seldom agree, as was first pointed out by the energetic German observer Johann Schröter in the late eighteenth century. Schröter made careful measurements of the time of dichotomy, when Venus appears as an exact half-disk. The results were surprising. When Venus is an evening object, and therefore waning, dichotomy is always early; during morning apparitions, when the phase is increasing, dichotomy is late. Moreover, the discrepancy seems to vary between one apparition and another; my own observations from 1934 to the present time indicate that the value is generally about three days. Undoubtedly the atmosphere of Venus is responsible; amateurs can do interesting work here.*

* When I wrote a monograph about Venus in 1956, I called this phenomenon 'the Schröter effect' – and the term now seems to have been accepted as official. It is certainly appropriate!

The atmosphere itself was first described in 1761 by M. V. Lomonosov, the first of Russia's famous astronomers. In that year Venus passed in transit across the face of the Sun, and Lomonosov saw that the limb was 'fuzzy' – from which he inferred, quite correctly, that there must be an atmosphere of considerable depth and density.

Transits of Venus are interesting to watch with the naked eye – or so I am told; I have never seen one, since the last occurred in 1882. Transits take place in pairs separated by eight years, after which there are no more for over a century. The transits of 1874 and 1882 will be followed by those of 2004 and 2012.

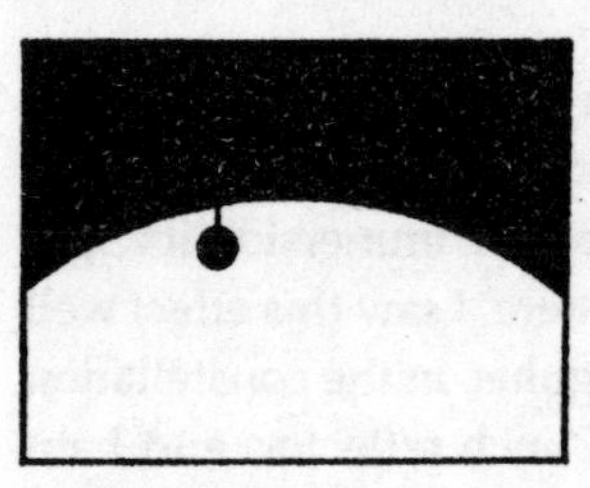

The Black Drop, seen at every transit of Venus.

During the seventeenth century Edmond Halley, the second Astronomer Royal, improved on an earlier suggestion by James Gregory that transits of Venus might be used to measure the length of the astronomical unit, or distance between the Earth and the Sun. For this, it was essential to time the exact moment when Venus passed on to the Sun's disk, and also to make observations from widely scattered points on the surface of the Earth. As the whole method is now completely obsolete there is no point in describing it further, but in any case the results were spoiled by an effect known as the Black Drop. When Venus passes on to the Sun, it seems to draw a strip of blackness after it; when this strip disappears, the transit has already begun. Here too the effect is caused by Venus' atmosphere, and nothing can be done about it. The 1874 and 1882 transits were well observed, but the results were disappointing, and since there are now much better ways of working out the Sun's distance the next pair of transits will not be regarded as of much importance. All the same, I am looking forward to 8 June 2004!

All sorts of stories are connected with transits of Venus. Consider, for instance, the French astronomer G. Legentil in 1761 and 1769. The first of these transits was expected to be well seen from India, and accordingly Legentil set sail for Pondicherry. Unluckily for him, the Seven Years War was raging, and he did not arrive until after the transit was over. Rather than risk a second delay, he decided to stay where he was for the next eight years, and observe the 1769 transit instead. Shortly before and shortly after the vital moment the sky was brilliantly clear, but the transit itself was completely hidden by clouds, and since it was rather too long for Legentil to wait until the next transit (that of 1874) he packed up and started for home. Twice he was shipwrecked, and finally reached Paris to learn that he had been presumed dead, so that his heirs were getting ready to distribute his property. The 1769 transit was also notable because Captain Cook's epic voyage to the South Seas was made to take astronomers to a suitable site for observing. Unlike Legentil, Cook and his associates were successful.

Very occasionally, Venus passes in front of a star and hides or occults it; when this happens the star fades and flickers for a few seconds before vanishing, because just before immersion its light is coming to us by way of Venus' atmosphere. I saw this effect well on 7 July 1959, when Venus occulted Regulus, in the constellation of Leo (the Lion). I watched it with a 12-inch reflector, and I am glad to have seen it, since it will be a very long time before Venus again passes in front of a bright star.

Look at Venus through a telescope – even a powerful one – and you will see little more than a bright disk. If you are lucky, you may be able to detect a few shadings, but the markings are always very vague, with indefinite outlines. They shift and change relatively quickly, so that there is no chance of their being surface features; they are clouds in the upper atmosphere of Venus, and on the whole they tell us remarkably little. Neither are ordinary photographs of much help, though pictures taken in ultraviolet do show some streaky features. It is hardly surprising that before 1962, when the first successful spacecraft passed by the planet, our ignorance of Venus as a world was more or less complete.

What we could do, however, was to study the upper atmosphere spectroscopically. In the 1930s it was established that the atmosphere is very different from ours, and is made up largely of the heavy gas carbon dioxide. Since this gas would be expected to sink rather than rise, it was reasonable to assume that the atmosphere was carbon dioxide all the way down to the surface, producing a 'greenhouse effect' and making Venus a very warm world indeed. But could there be seas there?

One man who believed so was the Swedish chemist Svante Arrhenius, whose work was good enough to win him a Nobel Prize. In 1918 he gave a vivid and attractive picture of Venus, which he assumed to be a world rather in the same state as the Earth used to be over 200 million years ago, in the Carboniferous Period, when the Coal Forests were being laid down and the most advanced life-forms were amphibians; even the great dinosaurs lay well in the future. According to Arrhenius:

> 'A very great part of the surface of Venus is no doubt covered with swamps, corresponding to those on the Earth in which the coal deposits were formed, except that they are about 30°C warmer. No dust is lifted high into the air to lend it a distinct colour; only the dazzling white reflex from the clouds reaches the outside space and gives the planet its remarkable, brilliantly white lustre. The powerful air-currents in the highest strata of the atmosphere equalize the temperature difference between equator and poles almost completely, so that a uniform climate exists all over the planet analogous to conditions on the Earth during its hottest periods.
>
> 'The temperature on Venus is not so high as to prevent a luxuriant vegetation. The constantly uniform climatic conditions which exist everywhere result in an entire absence of adaptation to changing exterior conditions. Only low forms of life are therefore represented, mostly, no doubt, belonging to the vegetable kingdom; and the organisms are of nearly the same kind all over the planet. The vegetative processes are greatly accelerated by the high temperature. Therefore, the lifetime of organisms is probably short. Their dead bodies, decaying rapidly if lying in the open air, will fill it with stifling gases; if embedded in the slime carried down by the rivers, they speedily turn into small lumps of coal, which later, under the pressure of new layers combined with high temperature, become particles of graphite . . . The temperature at the poles of Venus is probably somewhat lower, perhaps by 10°C, than the average temperature on the planet. The organisms there should have developed into higher forms than elsewhere, and progress and culture, if we

> may so express it, will gradually spread from the poles toward the equator. Later, the temperature will sink, the dense clouds and the gloom disperse, and some time, perhaps not before life on Earth has reverted to its simpler forms or even become extinct, a flora and a fauna will appear, similar in kind to those which now delight our human eye, and Venus will then indeed be the "Heavenly Queen" of Babylonian fame, not because of her radiant lustre alone, but as the dwelling-place of the highest beings in our Solar System.'

Certainly it is a fascinating picture, but there was also a suggestion that the planet could be a bone-dry dust desert, without a trace of moisture anywhere. Both ideas were still current in the 1950s, plus some rather unusual variations – such as Sir Fred Hoyle's suggestion that there might be oceans of oil, so that Venus 'is probably endowed beyond the dream of the richest Texas oil-king'. More credence was given to a theory by two eminent American astronomers, Fred Whipple and Donald Menzel, according to which the oceans were of ordinary water, with the clouds composed of equally ordinary H_2O.

In view of the limited information available at the time, the Whipple–Menzel marine theory was perfectly reasonable. Presumably the carbon dioxide in the atmosphere would have fouled the water, producing seas of soda-water (though I remember commenting at the time that the chances of finding any whisky to mix with it did not seem to be good). It seems likely that life on Earth began in our warm seas, and that at that epoch the atmosphere contained much more carbon dioxide and much less free oxygen than it does now, so why should not Venus be a world in a 'primitive' condition, capable of evolving along the same lines as the Earth has done and producing the same sort of advanced life? On this score, at least, Whipple and Menzel did not differ markedly from Arrhenius.

One point which could be tackled by Earth-based methods was that of the rotation period. The 'year' of Venus is almost 225 Earth days, but visual observations were unable to produce a reliable rotation period, though it was widely thought that the spin might be 'captured' in the same way as Mercury's was then thought to be. (One enthusiastic visual observer, Leo Brenner, made drawings from which he gave a period accurate to 1/1000 of a second!) The first positive information came in 1956, when spectroscopic studies indicated that the rotation period must be very long indeed. We now know it to be just over 243 Earth days, so that technically a Venus 'day' is longer than its 'year'. To make matters even stranger, Venus spins in a direction in a sense opposite to that of the Earth or Mars: east to west. If you could see the Sun from the planet's surface, it would rise in the west and set in the east 118 Earth days later.

Nobody knows why Venus behaves in this peculiar way. Suggestions that in its early history it was hit by a massive body and literally tipped over do not seem at all plausible, but it is difficult to think of any alternative. Moreover, we now know that the upper clouds have a rotation period of only four days, so that the whole situation is highly complicated. The four-day period was originally announced by French astronomers in the early 1960s, from studies of the vague shadings. I admit that I was highly sceptical – but they were right, and I was wrong.

Before coming on to Space Age developments, I must pause to say something about the Ashen Light, which has been known ever since the seventeenth century; it was first recorded by the Jesuit astronomer Giovanni Riccioli, best remembered today as the man who gave names to the craters of the Moon.

When the Moon is a crescent, high enough to be seen against a reasonably dark background, the unlit or 'night' part of the disk can generally be seen shining dimly. This is often termed 'the Old Moon in the Young Moon's arms', and there is no mystery about it; it is due to light reflected on to the Moon from the Earth. (The first man to explain this, incidentally, was none other than Leonardo da Vinci.) Telescopically a similar effect has been seen on Venus, but it cannot be due to the same sort of cause, because Venus has no moon. The Ashen Light has been recorded by almost every serious observer of the planet; for many years it was officially dismissed as a mere contrast effect, and it has never been properly photographed, but I have seen it so often, and so clearly, that I have absolutely no doubt about its reality. Perhaps the most intriguing explanation of it was due to Franz von Paula Gruithuisen, in the early nineteenth century. Gruithuisen was an enthusiastic observer who was concerned mainly with the Moon, but unfortunately his imagination was so vivid that he tended to bring a good deal of ridicule on himself – for instance, he was convinced that he had found an ancient city on the Moon with 'dark gigantic ramparts'. With regard to Venus, he pointed out that the Ashen Light had been observed in 1759 and again in 1806, an interval of 47 terrestrial or 76 Venus years. He went on to say:

> 'We assume that some (Venus) Alexander or Napoleon then attained universal power. If we estimate that the ordinary life of an inhabitant of Venus lasts 130 Venus years, which amounts to 80 Earth years, the reign of an Emperor of Venus might well last for 76 Venus years. The observed appearance is evidently the result of a general festival illumination in honour of the ascension of a new emperor to the throne of the planet.'

Later on, Gruithuisen modified this theory somewhat. Instead of a coronation ceremony, he suggested that the Light might be due solely to the burning of large stretches of jungle to produce new farm land, and added that 'large migrations of people would be prevented, so that possible wars would be avoided by abolishing the reason for them. Thus the race would be kept united.'

There are perhaps certain objections to this idea. Other theories involved phosphorescent seas, high-altitude auroræ, and exaggerated night-glow. Today it seems likely that electrical phenomena in the upper atmosphere of Venus are responsible, but amateur observations can be a real help in clearing the matter up – simply because the Ashen Light can be seen only when Venus is a crescent, and then only for very limited periods, so that the data are very incomplete.

The first Venus probe was launched by the Russians on 12 February 1961, but contact with it was lost at a fairly early stage, so that we will never know what happened to it. The next attempt, by the Americans, was even less helpful; Mariner 1, dispatched from Cape Canaveral on 22 July 1962, fell in the sea. Success came with Mariner 2, launched on 27 August of the same year. On 14 December it passed within 21,000 miles of Venus, and sent back a tremendous amount of information, much of which was frankly disappointing.

The long rotation period was confirmed; there was no sign of

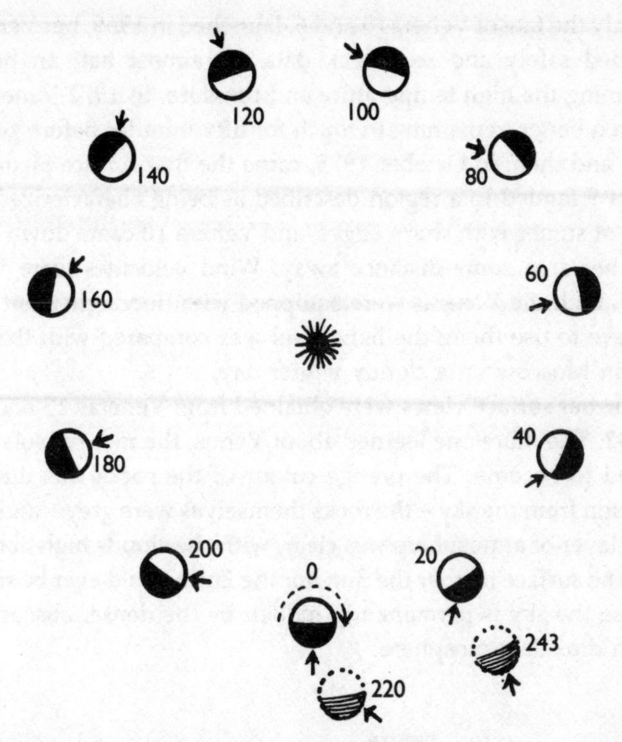

The rotation of Venus. The rotation period is 243 days, the orbital period 224.7 days. The arrow indicates one fixed point on the surface, which has a daylight period of about 59 days followed by an equal period of darkness.

any magnetic field, and temperature measurements showed that Venus really is scorching hot. The surface temperature is now known to be over 500 degrees Centigrade (900 degrees Fahrenheit), and the idea of wide oceans has had to be jettisoned; at that sort of temperature liquid water could not exist, despite the high atmospheric pressure. Mariner 2 showed that all hope of finding life on Venus must be given up.

There followed a long series of Russian attempts to carry out controlled landings, and obtain information direct from the planet's surface. For some time there were regular failures, either because of loss of contact or because – as we have since found – the pressure of the atmosphere had been seriously underestimated, and the probes were literally crushed on their way down. This was certainly the fate of Veneras 5 and 6, launched in 1969, but Venera 7 landed safely and sent back data for almost half an hour, confirming the high temperature and pressure. In 1972 Venera 8 did even better, remaining in touch for fifty minutes before going silent; and then, in October 1975, came the first surface pictures. Venera 9 landed in a region described as being characterized by heaps of stones with sharp edges, and Venera 10 came down in a smoother area some distance away. Wind velocities were very gentle. Both the Veneras were equipped with floodlights, but did not have to use them; the light-level was compared with that at noon in Moscow on a cloudy winter day.

Further surface views were obtained from Veneras 13 and 14 of 1982. The more one learned about Venus, the more desolate it seemed to become. The orange colour of the rocks was due to reflection from the sky – the rocks themselves were grey – and the lower layer of atmosphere was clear, with the clouds high above. From the surface neither the Sun nor the Earth could ever be seen, because the sky is permanently hidden by the dense, obscuring, carbon dioxide atmosphere.

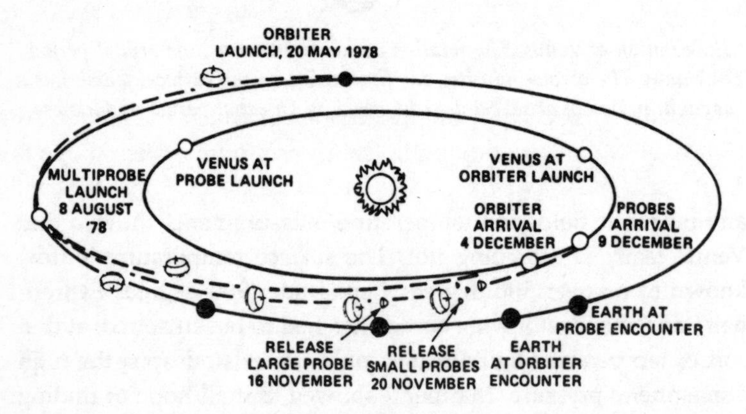

Trajectory of the Venus 'armada' of 1978. The orbiter continued transmitting until the autumn of 1992!

The American attack on Venus was different inasmuch as it concentrated on mapping the surface by radar, both from Earth (mainly with the huge radio telescope at Arecibo in Puerto Rico, built in a natural hollow in the ground) and from spacecraft. A whole armada was launched in 1978, made up of one orbiter and a 'bus', which was an entry probe carrying four smaller vehicles which came down at various points on the surface. The orbiter was particularly successful, and in fact did not come to the end of its career until 9 October 1992, when it finally lost contact.

There was an interesting encounter in June 1985, when the Russian Vega probes, on their way to a rendezvous with Halley's Comet, dropped balloons into Venus' atmosphere. These were monitored for some hours as they drifted around at various levels. Since then there have been the Galileo and Magellan missions. The Galileo space-craft was aimed at Jupiter, and its pass of Venus in February 1990 was merely an aside, but Magellan has been orbiting Venus ever since August 1990, and has sent back much the best radar pictures to date. It was still operating excellently in 1993.

By now almost ninety per cent of Venus has been mapped, and the results are truly fascinating. Venus is a volcanic world, and there seems every reason to believe that active vulcanism is going on now. Much of the surface is covered with a huge, rolling plain, and there are two main highland areas, Ishtar Terra in the north and Aphrodite Terra mainly in the southern hemisphere; Ishtar is about the size of the continental United States, while Aphrodite is decidedly larger. There are mountains, of which the highest, the Maxwell Mountains adjoining Ishtar, rise to five miles above the adjacent landscape; there are valleys, craters and features termed 'arachnoids' from their superficial resemblance to spiders' webs. Arachnoids seem to be circular volcanic structures surrounded by complex features of all kinds.

There are volcanoes, many of which seem to be of the shield type and not dissimilar to the volcanoes of Hawaii, though on a much grander scale. One highland area, Beta Regio, is dominated by two major shield volcanoes, Rhea Mons and Theia Mons, which are probably active; another active region – we think – is Atla Regio, bordering Aphrodite, where we find the huge volcano Sapas Mons, with a 250-mile base and a height of at least one mile. Lava-flows are evident here, as they are over much of Venus, and Sapas Mons has pits on its summit.

Quite apart from all this, we do at last have a good knowledge of the structure and composition of the atmosphere. Remember that the solid body of the planet rotates in 243 days, whereas the upper clouds have a period of only four days – a case of super-super-rotation. The upper winds are of hurricane force, but at the surface there is almost dead calm, which explains why the surface features are much less eroded than might have been expected. On Venus the pressure is about ninety times that of the Earth's air at sea-level, and the clouds are made up chiefly of sulphuric acid. There must be 'rain' but not of water; the Venus rain is of sulphuric acid droplets, which evaporate well before reaching the actual surface.

Venus has no detectable magnetic field, so that its heavy, iron-rich core must be smaller than that of the Earth, both relatively and absolutely. Above the core comes the mantle, and then the crust. On Earth, the crust slides around over the hot mantle, and this is why our volcanoes do not last for ever. A volcano builds up over a stationary hot spot in the mantle, and when the crustal slip carries the volcano away it ceases to erupt. This is what has happened on the Big Island of Hawaii. Mauna Kea, now the site of one of the world's great observatories, has been carried away from the hot spot and has become extinct; it has not erupted for thousands of years, and will never do so again. The hot spot now lies below its neighbour Mauna Loa, which is very active indeed. During the past forty years or so the science of 'plate tectonics' has become all-important to geologists, and it has been established that the Earth's crust is indeed made up of various well-defined plates, moving relative to each other. This does not seem to happen on Venus, so that when a volcano is formed it can remain over its hot spot for a very long period, building up to immense size.

Venus is named in honour of the Queen of Olympus; it was therefore decided that all the surface features should have female names – though there is one exception: the Maxwell Mountains were named in honour of the great Scottish mathematician, James Clerk Maxwell, before the official edict was confirmed. Many of the currently-accepted names are familiar; for example there are craters named Earhart, Nightingale, Pavlova, Colette and Lise Meitner, while other features honour Guinevere, Cleopatra, Atalanta and Helen. Recent additions have been a little less obvious; for example, have you ever heard of

Quetzalpetlatl, Hwangcini or Xiao Hong? (In case you are baffled, Quetzalpetlatl is an Aztec fertility goddess, Hwangcini an eighteenth-century Korean poetess, and Xiao Hong a Chinese novelist!)

If Venus and the Earth are near-twins, we have to ask ourselves why they are so different. The answer must surely lie in the fact that Venus is much closer to the Sun. It is thought that in the early history of the Solar System, over four and a half thousand million years ago, the Sun was less luminous than it is now, and Venus and the Earth began to evolve along similar lines, possibly developing the same sorts of atmospheres and oceans. But then the Sun became hotter, and so far as Venus was concerned the results were dire. Molecules of water vapour in its atmosphere were broken up by short-wave radiation from the Sun, and then re-combined to form molecules of oxygen and hydrogen; the lighter hydrogen rose to the top of the atmosphere and escaped into space, while the oxygen combined with the surface rocks. The net result was the loss of water, and before long, on the cosmical scale, Venus had become bone dry. The same process did not operate on the Earth, because of the lower temperature; most of the atmospheric water vapour was kept below an altitude of ten miles or so, where it was safe. Very little of it reached the uppermost layers.

As the process continued, Venus suffered a sort of runaway greenhouse effect. The carbonates were driven out of the rocks, and before long Venus changed from a potentially life-bearing world into the furnace-like inferno of today. So we are left with a planet where the atmospheric pressure is crushing, the temperature is intolerably high, and the clouds are rich in deadly acid. Go to Venus and step outside your spacecraft, and you will be promptly suffocated, fried, squashed and corroded. It is not a pleasant prospect.

Whether Venus will ever be visited by astronauts remains to be seen, but certainly there is no chance of it in the foreseeable future. There have been suggestions of 'terraforming' the planet, breaking up the carbon dioxide molecules in the atmosphere and releasing free oxygen, but anything of the sort is so completely beyond our present technology that to discuss it further is really rather pointless. To us, Venus is a world to be viewed from a respectful distance – so what can the telescopic observer usefully do?

Features to note are: the phase (remembering that observation and theory usually disagree to some extent), any visible shadings, any irregularities in the terminator, and any sign of the Ashen Light. Filters are often a great help, and for the Ashen Light I recommend a special occulting eyepiece of the type described on page 63. Obviously the Ashen Light can be seen only when the crescent Venus is reasonably high, but for most other observations the best views are likely to be obtained in daylight – which means using a telescope equipped with setting circles.

In some ways it must be admitted that Venus has been something of an anti-climax. Little more than a century ago Camille Flammarion could still write: 'Organized life on Venus must be little different from terrestrial life . . . this world differs little from ours in volume, weight, density, and in the duration of its days and nights. It should then be inhabited by vegetable, animal and human races little different from those which people our planet.' Alas, Venus has proved to be probably more hostile than any other world in the Solar System. It may have been named after the Goddess of Love and Beauty, but conditions on its surface are much more akin to the conventional idea of hell.

CHAPTER SEVEN

The Earth

When writing a general book about the Solar System, it is not easy to decide just how to deal with the Earth. It is a normal planet, and we regard it as exceptional only because we happen to live on it, but it is the concern of geophysicists rather than astronomers, so that for the moment I feel that it will be best for me to limit myself to matters which have a purely astronomical slant.

There is nothing unusual about the Earth's orbit. Our average distance from the Sun is 92,975,000 miles; the sidereal period is 365¼ days, and the mean orbital velocity is 18½ miles per second, or 66,000 mph. The Earth's path round the Sun is not perfectly circular; we reach perihelion in January, aphelion in July. The seasons are not due to the changing distance (91½ to 94½ millon miles) but to the tilt of the Earth's axis, which amounts to 23½ degrees to the perpendicular to the orbital plane. In position 1 in the diagram, the northern hemisphere is tilted sunward, and Europe has its summer; in position 2, it is northern winter. The fact that the Earth is three million miles closer to the Sun in position 2

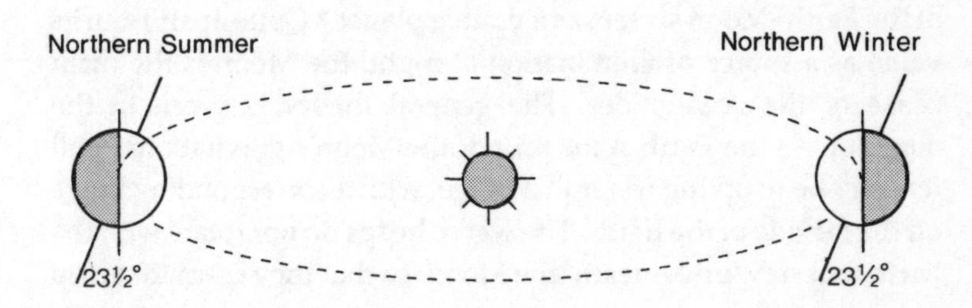

The Seasons. In position 1, the north pole is tilted toward the Sun (northern summer), in position 2 it is southern summer.

makes very little difference, and the effects on the world climate are more or less masked by the unequal distribution of land and sea in the two hemispheres. Of the other planets, Mars, Saturn and Neptune have axial tilts of the same general type as ours, while Jupiter and Mercury are almost 'upright'; Venus, as we have seen, is an oddity – it rotates in the sense opposite to ours – and Uranus is even more curious, because it is tilted at more than a right angle.

In size and mass the Earth is equally unremarkable. It has a diameter of 7926 miles as measured through the equator, but only 7900 miles as measured through the poles; the globe is appreciably flattened, rather less than with Mars but more so than with Mercury or Venus. The specific gravity is 5.5, so that the Earth 'weighs' 5.5 times as much as an equal volume of water would do. Mercury is almost equally dense; Venus and Mars rather less.

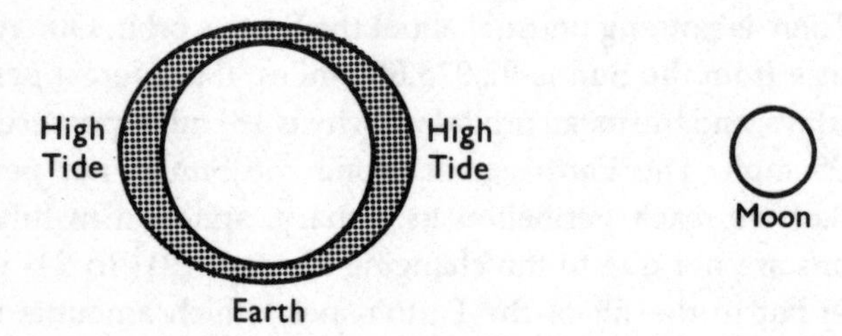

The tides. The diagram is not to scale, and of course the depth of the water-shell is wildly exaggerated.

In only one respect is the Earth unique: it is the only relatively small planet to be attended by a large satellite, and I always think of the Earth–Moon system as a double planet.* Quite apart from its value as a source of illumination at night, the Moon is the main cause of the ocean tides. The general theory is given in the diagram. As the Earth spins round, the Moon's gravitational pull tends to heap up the water in a bulge, with a corresponding bulge on the far side of the Earth. The water-heaps do not rotate with the Earth, but stay 'underneath' the Moon, so that they seem to sweep

* I am not counting Pluto and its attendant Charon, because there now seems no doubt that Pluto is not a proper planet.

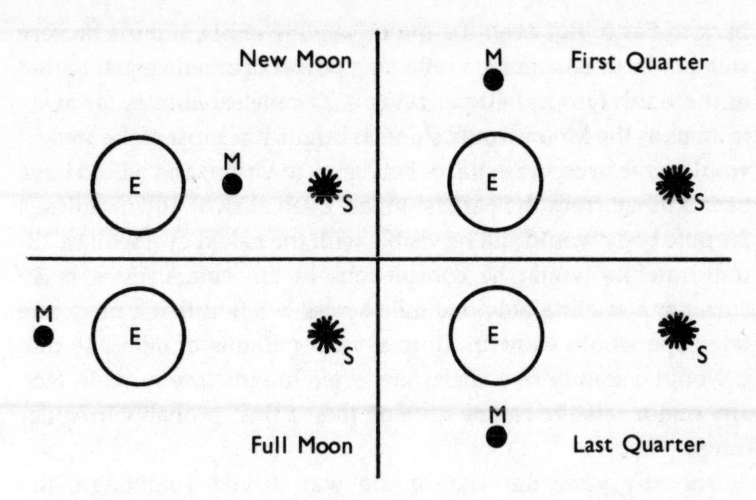

(Left) *Spring tides: the Sun and Moon are pulling in the same sense.* (Right) *Neap tides: the Sun and Moon are pulling at right angles to each other.*

right round the world twice a day; since there are two heaps, each point on the Earth's surface has two daily high tides. In practice there are many complications, and the Sun also has a strong tide-raising effect; when the Sun and Moon are pulling in the same sense (that is to say, at new or full moon) the tides are at their strongest. These tides are known as spring tides, which is a misleading term inasmuch as it has nothing to do with the season of spring. The weakest tides, at half moon, are called neap tides.

Tides on other planets would be of different type even if any of the other planets had oceans on their surfaces. Venus has no satellite, while any hypothetical Martian seas would be calm and sluggish, since Mars is much further away from the Sun and its two dwarf moons, Phobos and Deimos, are much too puny to be potential tide-raisers. In fact it seems likely that both these tiny bodies are ex-asteroids which were captured by Mars long ago. If so, can there be any minor satellites of the Earth?

The idea of a second satellite is not new. Jules Verne used it in his famous novel *From the Earth to the Moon;* it was an essential part of his plot, since the wanderer pulled the man-carrying projectile out of its original path and swung it round the Moon back to Earth. But even if a minor satellite exists, it must be very small indeed. Assuming a reflecting power or albedo equal to that of the Earth (around 40 per cent), a 25-mile satellite as far away from us as the Moon would shine as brightly as most of the stars; it would have been the equal of Betelgeux in Orion, and would have been known from the earliest times. Even at two million miles, a 25-mile body would still be visible with the naked eye, while a 12-mile satellite would be conspicuous in binoculars. Even if we consider a satellite only one mile across, we find that a moderate telescope would show it out to several millions of miles, so that it would certainly have been identified long before now. In fact, any minor satellite can be nothing than a tiny, probably irregular lump.

Shortly after the end of the war, Clyde Tombaugh, the discoverer of Pluto, carried out a long and systematic hunt for minor satellites. His equipment would have been capable of picking up a football-sized body a thousand miles away, even if the object had been made of dark rock, while a 10-foot diameter satellite would have been detectable out to 10,000 miles. Nothing whatsoever was found.

Some years ago there was an interesting speculation about an asteroid – No. 1685, Toro – which made a fairly close approach to us on 8 August 1972, passing by at only 13,000,000 miles. Its diameter is 6 miles. Its orbit is not so very unlike that of the Earth, and it approaches us periodically, so that there were press claims that it ranked as an Earth satellite. Of course it is nothing of the kind; it is a perfectly normal asteroid.

There have also been suggestions that there may be loose clouds of meteoritic material moving round the Earth in the same orbit as that of the Moon, one cloud keeping 60 degrees ahead of the Moon and the other 60 degrees behind. These 'stable' points are known as Lagrangian points in honour of the great French mathematician. There is nothing impossible in this – we will come to Lagrangian points again when discussing the Trojan asteroids – and the Polish astronomer K. Kordylewski has claimed that the clouds have been actually seen; but even if they exist they must be of very low mass, and I admit to being sceptical. I will change my mind only when somebody manages to photograph them.

At least the presence of interplanetary material makes itself

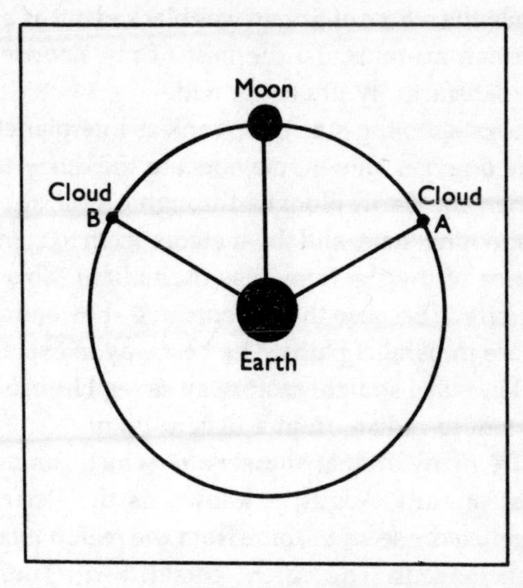

Lagrangian points, 60 degrees away from the Moon but in the same orbit. These are the positions of the unconfirmed Kordylewski clouds.

obvious with the sky-glows known as the Zodiacal Light and the Gegenschein. The Zodiacal Light extends along the ecliptic, and can be seen for only a short period either after sunset or before sunrise. From Britain it is never conspicuous, but from countries with clearer skies and less light pollution it can be quite spectacular, and can even become brighter than average parts of the Milky Way. It is due to particles in the Solar System scattered along and near the main plane; the average diameter of a particle is one or two microns. (One micron is equal to one-millionth of a metre.) Since the Zodiacal Light extends along the ecliptic, it is best seen when the ecliptic is most nearly vertical to the horizon, i.e. in February/March and again in September/October. It is interesting to photograph, but you need a fairly fast film, and an exposure of from 5 to 30 minutes is to be recommended.

The Gegenschein is much less easy to see. It takes the form of a dim glow exactly opposite to the Sun in the sky; its greatest diameter may be as much as forty times that of the full moon. The German name is translated as 'Counterglow'. It too is caused by interplanetary material. I have seen it well only once, and that was in 1940, when the whole of Britain was blacked out as a precaution against German air-raids. To the best of my knowledge, it has never been satisfactorily photographed.

Meteors, or shooting-stars, also rank as interplanetary débris, but here the origin is known: meteors are the dusty trails left by comets. When the Earth ploughs through a trail, the result is a shower of shooting-stars, and the meteors seem to come from one particular area of the sky known as the radiant. This is purely a perspective effect, because the meteors in a shower are travelling through space in parallel paths. The best way to explain this is to look at the lanes of a straight motorway as seen from a bridge; the lines will seem to radiate from a distant point.

There are many annual showers, of which the most spectacular is that of early August – known as the Perseid shower, because the meteors seem to come from the region marked by the constellation Perseus. The parent comet, Swift-Tuttle,* has an orbital period of 130 years, and made its last approach in 1992. When a meteor dashes into the upper air it sets up friction against the atmospheric particles, and burns away. (Note that what we see is not the tiny particle itself, but its effects upon the atmosphere.) Of course, no meteor can be seen unless the air around it is dense enough to produce heat by friction, even bearing in mind that a meteor may enter at anything up to 45 miles per second. In general it has been found that a meteor becomes luminous at a height of about 120 miles above sea-level, and will burn out by the time it has fallen to 40 miles, ending its groundward journey in the form of ultra-fine dust. Quite apart from the known showers there are also sporadic meteors, not connected with any known comet, which may appear from any direction at any moment.

There can be no meteors seen from the airless Moon, because there is nothing to set up friction and make the incoming particle glow. From Venus, a meteor would be destroyed long before it could penetrate far. If you want to see meteors, you must either stay at home or else go to Mars, where meteor displays are presumably quite common.

Meteorites are very different. They are not connected with

* So named because it was discovered, at its 1862 return to the Sun, independently by Lewis Swift and Horace Tuttle.

comets, or with shooting-star meteors; they come from the asteroid belt, and there may indeed be no distinction between a large meteorite and a small asteroid. When a meteorite lands, it may produce a crater. One splendid example is the Meteor Crater in Arizona (it really should be called the Meteor*ite* Crater), which is almost a mile wide; it was formed well over 20,000 years ago, and is well worth visiting. Unquestionably there have been many major impacts in the past, and there is a very serious suggestion that one such collision, around 65,000,000 years ago, caused such a dramatic change in our climate that the dinosaurs, which had been lords of the world for so long, were unable to adapt to the new conditions, and died out. I will have more to say about this later.

Not many people have seen a meteorite fall, but there were many observers of the Barwell Meteorite of Christmas Eve, 1965, which blazed across the English sky before breaking up and scattering fragments in Leicestershire. More recently there has been the Bovedy Meteorite, which also was widely seen during its flight; though most of it fell in the Irish Sea, many fragments of it were found in Northern Ireland. (I missed it by about a minute. I had been in my observatory in Selsey, and had just gone indoors to check some charts when the meteorite shot over.) The latest British meteorite fell at Glatton, Cambridgeshire, on 5 May 1991; it weighed only 767 grams, and came down some sixty feet away from a Mr Pettifor, who had been doing some casual gardening. Incidentally, there is no authentic record of anyone having been killed or badly injured by a tumbling meteorite, though it is true that one or two people have had narrow escapes.

Most museums have meteorite collections, but if you want to see the holder of the heavyweight record you must go to Hoba West, near Grootfontein in Southern Africa. The meteorite is still lying where it fell in prehistoric times; I doubt whether anyone will try to run away with it, since its total weight is well over sixty tons.

Before the Space Age, meteorites represented the only extra-terrestrial material which we could actually handle. Analysis shows that there are two main types – stones and irons – though there are many subdivisions. It has even been claimed that some meteorites have been blasted away from Mars or the Moon, though this is something about which I am highly suspicious.

Neither have I much faith in Sir Fred Hoyle's theory that life on Earth was brought here by way of a meteorite. True, we are very uncertain about the origin of life, but to me at least the meteorite theory seems to raise far more problems than it solves.

Come now to atmospheric phenomena, as distinct from interplanetary débris. Remember that the atmosphere is made up of several layers; the nomenclature is quite complicated, so I propose to simplify it and use only the most basic terms.

Most of the atmosphere is made up of two gases, nitrogen (78 per cent) and oxygen (21 per cent), with much smaller amounts of other gases such as argon and carbon dioxide, plus a variable amount of water vapour. No other planet in the Solar System has an atmosphere like ours. Titan, the largest satellite of Saturn, has an atmosphere rich in nitrogen, but most of the rest is methane (marsh gas), and there is almost no free oxygen.

The lowest part of the atmosphere is known as the troposphere. It extends upward for from about 5 miles to over 11 miles according to latitude (it is deepest over the equator), and it is here that we find all our normal clouds and weather. The temperature falls with increasing height, and at the top of the troposphere it has dropped to about −44 degrees Centigrade (−80 degrees Fahrenheit). At this altitude, of course, the density is very low.

Above the troposphere comes the stratosphere, which extends up to about 30 miles. Surprisingly, the temperature does not go on falling; it actually rises, reaching +15 degrees Centigrade (+60 degrees Fahrenheit) at the top of the layer. This is because of the presence of ozone, which is a special form of oxygen; an ozone molecule is made up of three oxygen atoms (O_3) instead of the usual two. Short-wave radiation from the Sun warms the ozone layer, and prevents the temperature in the stratosphere from falling any further. Yet remember that there is a difference between scientific 'temperature' and what we normally understand by 'heat'. Temperature depends upon the rate at which the atoms and molecules move around; the quicker the movements, the higher the temperature. But in the stratosphere there are so few molecules left that the 'heat' is negligible – and there is some analogy here with a firework sparkler; every spark is at a high temperature, but is of such low mass there is no danger in holding the firework in your hand. (A glowing poker is 'only' red-hot, and therefore at a much lower temperature, but I do not recommend grasping one.)

Much has been heard recently about 'holes' in the ozone layer, which have been attributed to our own activities in sending ozone-harmful materials high into the air. If this is true, then we must be on the alert, because the ozone layer shields us from dangerous radiations coming from space.

Above the stratosphere comes the ionosphere, stretching upwards for several hundreds of miles.* It is here that we find the strange noctilucent clouds, which lie at an altitude of around 50 miles and are quite unlike our familiar clouds. They may be due to ice crystals forming along meteor trails, though we are still not sure; at any rate, they are interesting to identify and to photograph. In the northern hemisphere they are best seen from May till August between latitudes 50 and 70 degrees; they show a marked ripple or wave structure, with veins, bands and whirls.

Meteors burn away in the ionosphere, which also acts as a screen against cosmic rays – which are not rays at all, but atomic particles coming in from space from all directions all the time. And ionospheric layers reflect some radio-waves back to the ground, making long-range communication possible. Finally, it is here that we find the lovely glows which we call auroræ, or polar lights.

Auroræ (aurora borealis in the northern hemisphere, aurora australis in the southern) are due to electrified particles sent out by the Sun. These particles collide with atoms and molecules in the upper air, and make it glow; because the particles are charged, they tend to spiral downward towards the magnetic poles, and so auroræ are best seen from high latitudes. From places such as Alaska and North Norway they are visible on almost every clear night; they are common in Scotland, but much less so in Southern England, though there are occasional brilliant displays (as on 13 March 1989 and 8–9 November 1991). Obviously, auroræ are most common when the Sun is at its most active, and the solar wind is at its 'gustiest'.

* The ionosphere is often subdivided into the mesosphere (up to 50 miles) and the thermosphere (to 180 miles); in the thermosphere the temperature rises to 1900 degrees Centigrade (3500 degrees Fahrenheit), but of course the air density is so low that there is virtually no 'heat'.

Generally speaking, the sharp lower boundary of an aurora lies 60 miles above sea-level, with maximum activity at 70 miles and the normal upper boundary at 190 miles, though in exceptional cases the maximum altitude may reach as much as 600 miles. Displays may take various forms; mere glows, or arcs, rays, bands, draperies or curtains. There may be vivid colours and quick movement. The best instrument for observing them is the naked eye – plus a good camera! Many reports have come in of crackling sounds and acrid odours, but I have to confess that I am somewhat sceptical about noisy auroræ and decidedly so about smelly auroræ.

There can be no auroræ on Mercury or the Moon. Mars may show them, but we have no proof; and with regard to Venus there just may be some link between auroral-type phenomena and the Ashen Light. However, the outer giant planets produce strong auroræ, though with Uranus and Neptune the main displays take place closer to the equator than the poles of the globes, because of the sharp tilt between the rotational and magnetic axes.

Above the ionosphere comes the exosphere. This is the outermost part of the mantle of air, and has no definite boundary; it simply tails off until the density is no greater than that of the general atmospheric medium. It is what is termed a collisionless gas, so that the atoms and molecules simply orbit the globe without suffering constant hits from their neighbours.

Next let us turn to the magnetosphere, which may be defined as the region in which the Earth's magnetic field is dominant. It is shaped rather like a tear-drop, with the tail pointing away from the Sun. On the sunward side of the Earth it extends for only about 40,000 miles, but on the dark side it reaches out much further. The Sun sends out a constant stream of particles, making up the solar wind. When these solar wind particles come towards the Earth they meet the magnetic field, and the result is a shock-wave.

Inside the magnetosphere there are two zones of intense radiation, known as the Van Allen zones in honour of James Van Allen, the American scientist who was principally responsible for their discovery. They were detected by instruments carried aboard Explorer 1, the first successful US artificial satellite, which was launched on 1 February 1958. There are two main belts, one with

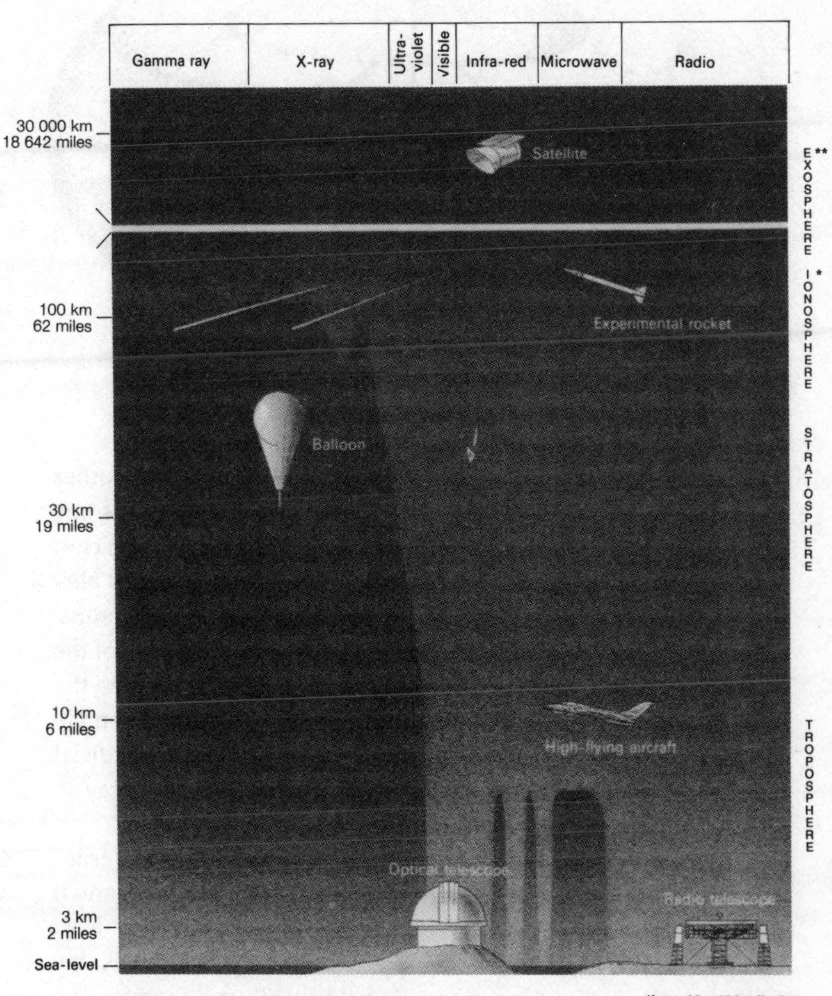

Cross-section of the atmosphere.

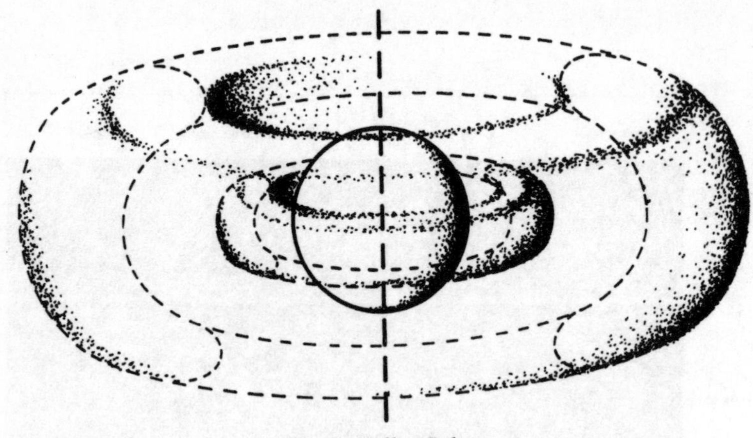

The Van Allen Belts

its lower limit at an altitude of just under 5000 miles and the other extending up to over 23,000 miles. (I must add that there is some doubt about the precise origin of the lower-belt particles; interactions between upper-air particles and cosmic rays may play an important rôle.) The inner belt, made up chiefly of protons, actually dips down towards the Earth's surface in the region of the coast of Brazil, because the Earth's magnetic field is offset from the axis of rotation, and this so-called South Atlantic Anomaly presents a distinct hazard to equipment carried in artificial satellites. Sensitive instruments become saturated if they stay in the danger-zone for too long.

The Earth's magnetic field is due to movements in the iron-rich core of the globe, but we cannot pretend that we understand it as completely as we would like to do. However, we can at least compare it with the fields of other planets. So far as we can tell, the Moon and Venus have no magnetic fields at all, and therefore no Van Allen-type zones; if there is a general magnetic field associated with Mars, it must be very weak indeed. But all the giants are powerful magnets, and in the case of Jupiter the encircling radiation zones are so strong that they alone would prevent any manned expedition from going there – even if Jupiter were a suitable target in other ways, which it certainly is not.

Most of our knowledge of the Earth's interior has been drawn from studies of the waves set up by earthquake shocks. This is no place to discuss them in any detail, but basically there are two types of earthquakes waves which concern us here. One type can travel through a liquid, while the other cannot. Our measurements of the size of the liquid core depend upon finding out just where the waves of the second type are stopped.

The Earth's crust has an average depth of 6 miles below the oceans, but 30 miles below the continents. Below comes the mantle, which extends down to 1800 miles and contains 67 per cent of the Earth's mass; partial melting of mantle material produces basalt, which issues from volcanic vents on the sea-bed. Underneath the mantle comes the liquid core, while the innermost part of the core is presumably solid. The central temperature is probably about 4000 degrees Centigrade (7000 degrees Fahrenheit), which is greater than for any other inner planet or for the Moon.

Geology can tell us a great deal about the past history of the Earth. We are confident about its age; we believe that the original atmosphere was lost, and that the present atmosphere was produced by gases and vapours sent out from the interior. Life, in its most primitive form, began at an early stage in world history, presumably in the seas. At first the 'new' atmosphere was very rich in carbon dioxide, and only when plants spread on to the lands did the situation change; plants removed the carbon dioxide and replaced it with free oxygen, by the process we call photosynthesis. If we could board Dr Who's time machine and project ourselves back to, say, the Cambrian Period, 500 million years ago, we would choke.

There have been periodical ice ages, whose cause is by no means certain; the last of these ended a mere 10,000 years ago, and there will no doubt be others in the future. All sorts of theories have been proposed, ranging from asteroidal impacts to changes in the shape of the Earth's orbit, but all things considered it is surely reasonable to suppose that the Sun is involved. After all, to a very minor extent the Sun is a variable star.

From the Moon, the Earth is a glorious sight, as the Apollo astronauts have found. From Venus it would never be seen – the dense cloud cover would hide it – though from just above the cloud-deck it would shine as a body of magnitude −6.5, and would be truly magnificent. From Mars it would be an inferior planet, showing lunar-type phases and moving very much as Venus seems to do to us. An observer as far out as Jupiter would have difficulty in seeing the Earth at all, and from the more remote planets it would be lost in the Sun's glare. We have to concede that it is of minor importance in the Solar System; but it is our home, and it suits us well.

CHAPTER EIGHT

The Moon

'That's one small step for a man; one giant leap for mankind.' Nobody will ever forget Neil Armstrong's words as he stepped out on to the barren rocks of the lunar Sea of Tranquillity. And there was, I remember, a general feeling that at last we had solved all the Moon's mysteries. Far from it – the Moon is still full of surprises.

To us it appears the most splendid object in the sky apart from the Sun, and it is not in the least surprising that lunar worship was common in ancient times – indeed, in some countries it lingers on even now. Yet we know that the Moon owes its eminence to the fact that it is so close to us. At its mean distance of less than a quarter of a million miles, it is about a hundred times nearer to the Earth than Venus can ever be.

I have said that in my view, at least, the Earth–Moon system should be regarded as a double planet rather than as a planet and a satellite. The Earth has 81 times the mass of the Moon, but with all other planet–satellite systems the discrepancy is much greater (as usual, I exclude Pluto). Titan, the senior member of Saturn's family, has only 1/4150 the mass of Saturn itself, while Jupiter's satellite Ganymede, which is actually larger than the planet Mercury, is even more puny relative to its primary. Therefore our Moon is exceptional, and we are entitled to regard it as a special case.

There has been a great deal of discussion about the origin of the Moon, and it cannot be said that the problem has been solved even yet. At least we have something to guide us, because analyses of the samples brought back by the Apollo astronauts and the Russian unmanned probes confirm that the Earth and the Moon are of approximately the same age. We also know that the Moon is less dense than the Earth, and 'weighs' only 3.3 times as much as

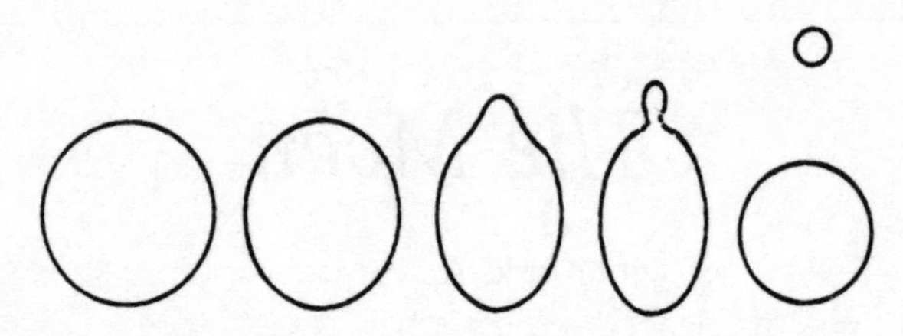

Tidal theory of the origin of the Moon. Unfortunately, it is mathematically untenable, and has been abandoned.

an equivalent volume of water would do, as against 5.5 times for the Earth. The mean density of the Moon is about the same as that of the Earth's mantle, and there is a much smaller heavy core – which is no surprise, particularly in view of the lack of an overall magnetic field.

The first widely-accepted theory was due to Sir George Darwin, son of the great naturalist. Darwin started by assuming that the Earth and the Moon originally formed one body, and that the Moon was thrown off as a fluid mass. In a modified version of this idea, the Earth had cooled down sufficiently to form a thin crust before the separation took place, and the sequence of events was worked out in considerable detail. The Earth, rotating rapidly on its axis, was in the state known as 'unstable equilibrium', so that it became egg-shaped, spinning about its shorter axis. Two main forces were acting upon it: the tides raised in it by the Sun, and its own natural period of vibration. When these two forces were in resonance (that it to say, acting together) the tides increased to such an extent that the whole body became first pear-shaped and then dumb-bell-shaped, with one 'bell' (the Earth) much larger than the other (the future Moon). Eventually the neck of the dumb-bell broke altogether, and the Moon moved away, settling into a stable orbit.

A strong supporter of the fission theory was W. H. Pickering, an American astronomer whose main interest was in lunar work. Pickering, who died as recently as 1938, went even further than Darwin had done, and wrote that if the theory were correct the thin crust of the otherwise fluid Earth must have been torn apart, leaving a huge hollow where the thrown-off mass had once been. He suggested that the hollow now filled by the Pacific Ocean was the scar left by the departing Moon, and that the shock caused by the final fracture was violent enough to crack the fragile crust in other places as well. It all sounded very plausible, but unfortunately it was then found that from a mathematical viewpoint the process simply will not work. A piece of material the size of the Moon could not be thrown off as Darwin believed, and in any case it could never remain as a separate body; it would break up, and most or all of it would fall back on to the Earth.

Next came the idea of the Moon as a formerly independent planet, which was captured when it happened to wander too close to the Earth – or perhaps the Earth and the Moon were born at the same time in the same region of space, so that they have always remained together. There were difficulties here too, because it would involve a set of very special circumstances; otherwise the Moon would either hit the Earth full-on, or would rush past it and remain independent.

The latest theory was proposed in 1974 by W. Hartmann and D. R. Davis. This time there is a collision between the Earth and a large impactor possibly about the size of Mars. The cores of the Earth and the impactor merged, and mantle débris ejected during the collision formed a temporary cloud of material round the Earth which subsequently collected together to form the Moon. This would certainly explain why the Moon's overall density is less than that of the Earth. Problems remain, but at least the theory does look promising.

It is often said that 'the Moon revolves round the Earth'. In a way this is true enough, but it does not tell us the whole story. Strictly speaking, the Earth and the Moon revolve round their common centre of gravity, much as the two balls of a dumb-bell will do when twisted by their joining arm. Since the Earth is 81 times as massive as the Moon, this centre of gravity or *barycentre* is displaced toward the Earth; in fact it actually lies inside the terrestrial globe, so that the simple original statement is good enough for most purposes so long as we remember that the mass of the Moon is far from negligible.

Everyone is familiar with the phases of the Moon. They must have been known from the earliest days of human history, and

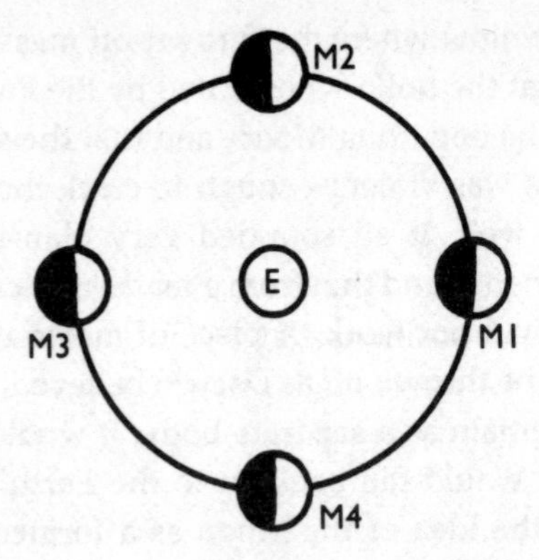

Phases of the Moon. New: M1. First Quarter: M2. Full: M3. Last Quarter: M4. The sunlight is assumed to be coming from the right.

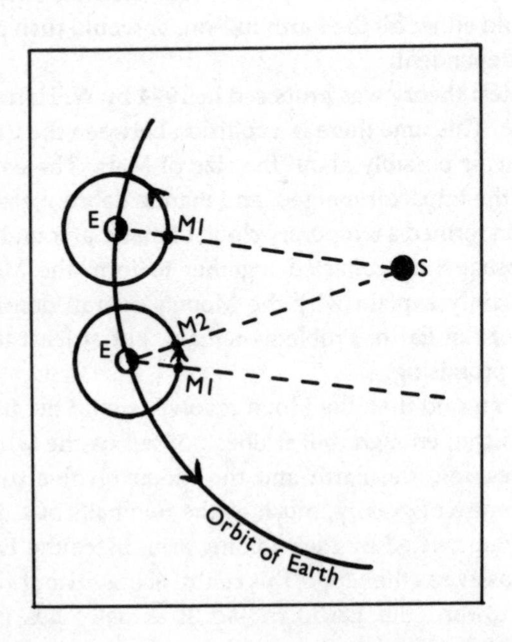

The interval between successive new moons is longer than the Moon's orbital period of 27.3 days. In the upper diagram the Moon is new at M1; but in the lower diagram it is not new until it has passed M1 (after one full circuit of the Earth) and has reached M2, because in the interval the Earth has moved on round the Sun.

there is no mystery about them once we realize that the Moon depends entirely upon reflected sunlight. The diagram given opposite is self-explanatory; the Moon is new in position 1, half at 2, full at 3, and half again at 4. Between 1 and 2, and again between 4 and 1, it is a crescent; between 2 and 3, and between 3 and 4, it is gibbous. The only points worth adding are that the true new moon is invisible, and that positions 2 and 4 are known as First Quarter and Last Quarter respectively.

The sidereal month, or time taken for the Moon to complete one journey round the Earth (or, more accurately, round the barycentre), is 27.3 days. However, the interval between successive new moons is rather longer, because the Earth itself is moving round the Sun. In the next diagram, the Moon is represented by M, the Earth by E and the Sun by S. In the upper position the Moon is new at M1, since it lies between the Earth and the Sun. After 27.3 days the Moon has arrived back at M1, but it is not now lined up with the Sun, because the Earth has moved along in its orbit. Only when the Moon has arrived at position M2 will it again be new. The *synodic month*, or interval between one new moon and the next (or one full moon and the next) is therefore 29.5 days instead of only 27.3.

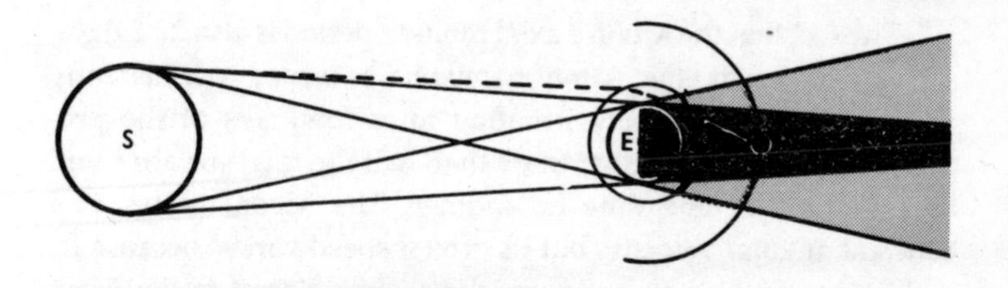

Theory of a lunar eclipse. The Moon passes into the cone of the Earth's shadow (black) during an eclipse, but some sunlight is refracted on to it through the Earth's atmosphere, as shown by the dashed line. The penumbra (shaded area) lies to either side of the main cone or umbra. Of course the Moon has to pass through the penumbra before reaching the umbra, and some eclipses are only penumbral – and none too easy to notice with the naked eye.

When full, the Moon sometimes passes into the shadow cast by the Earth. The result is a lunar eclipse; all direct sunlight is cut off from the Moon's surface, and the Moon becomes very dim until it emerges from the cone of shadow. It does not (usually) disappear completely, because some of the Sun's rays are bent on to it after having passed through the atmosphere of the Earth, but some eclipses are darker than others; it all depends upon conditions in our upper air. For example, the eclipse of December 1992 was very dark indeed, because of the amount of dust and ash sent into the high atmosphere by the violent volcanic eruption of Mount Pinotubo in the Philippine Islands.

Lunar eclipses may be either total or partial, and are more often seen than eclipses of the Sun. In fact, solar and lunar eclipses are about equally numerous, but a solar eclipse is visible from only a narrow strip across the Earth's surface, whereas a lunar eclipse can be seen from anywhere from which the Moon is above the horizon. Eclipses do not happen every month, because, as we have seen, the Moon's orbit is appreciably tilted.

It cannot be said that eclipses of the Moon are scientifically important, but they are always worth watching, and there are often vivid colours. As the Earth's shadow sweeps across the Moon the temperature on the lunar surface drops sharply, but some special areas cool down much less quickly than the average; these are known, rather misleadingly, as hot spots. One such region is the majestic crater Tycho.

The fact that the Moon's axial rotation period is also 27.3 days, so that it keeps the same hemisphere turned permanently Earthward, was intensely irritating to astronomers of the pre-Space Age. Actually, rather more than half the total surface could be examined at one time or another. The Moon rotates at a constant angular velocity, but its orbital speed varies, because its path is not circular; it moves quickest when closest to the Earth (perigee) and slowest when furthest away (apogee). Therefore, the position in orbit and the amount of axial spin become periodically out of step. We can see alternately first beyond one mean limb and then beyond the other, so that the Moon seems to rock very slowly to and fro; this 'libration in longitude' is evident over a period of a few days. There is also a 'libration in latitude', because the Moon's orbit is inclined to ours at an angle of five degrees, and we can see for some distance between alternate poles. Finally there is 'diurnal libration', due to our being placed on the surface of the Earth almost 4000 miles away from the centre of the globe; at moonrise

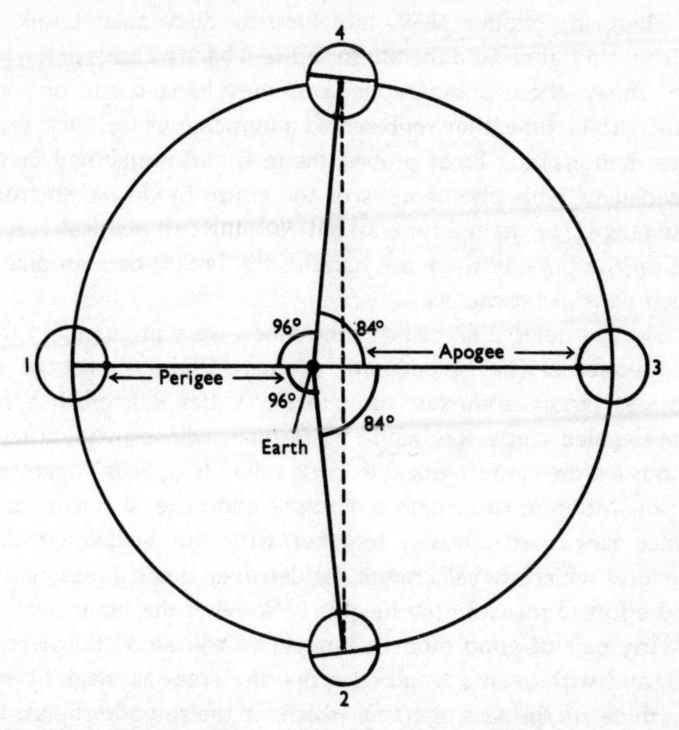

Libration in longitude. Because the Moon's orbit is not circular, its orbital velocity varies; it moves from 1 to 2, and from 4 to 1, more quickly than from 2 to 3 or 3 to 4, whereas its rotational velocity is constant. Therefore, the apparent centre of the visible disk as seen from Earth is displaced, and we can see for a little way beyond alternate mean limbs.

we can peer a little way beyond the mean limb. The overall result of these librations is that from Earth we can see a grand total of 59 per cent of the Moon's surface, though, of course, never more than 50 per cent at any one time.

The remaining 41 per cent is permanently out of view, and in the past there were some strange theories about it. Andreas Hansen, of Denmark, even believed that the Moon might be lopsided in mass, and that all the air and water had been drawn round to the far side, which might well be inhabited! He met with little support, but right up to 1959 it was still thought possible that the hidden regions might be very different from those which we have always known.

Then, in October 1959, the Russians dispatched Lunik 3, which went right round the Moon and sent back pictures of the far side. Today, these photographs seem very blurred and of poor quality; at the time, they represented a tremendous technical feat. Since then orbiting lunar probes, manned and unmanned, have provided us with photographs of the entire Moon, taken from close range. The main difference between the two hemispheres is that on the far side there are none of the broad, darkish plains which we still miscall seas.

The first telescopic maps of the Moon were produced in the first decade of the seventeenth century. Priority must go to Thomas Harriot, sometime tutor to Sir Walter Raleigh, but the most detailed study was made by Galileo, whose map showed various features in recognizable form. It had long been suggested that the Moon must contain mountains and valleys; Harriot and Galileo saw them clearly, together with the walled circular structures which we call craters. Galileo even made a reasonably good effort to measure the heights of some of the lunar peaks.

Any pair of good modern binoculars will show the details well, and with even a small telescope the scene is magnificent. First there are the seas or *maria*, which are the smoothest parts of the surface even though by ordinary standards they are still extremely rough. They have been given romantic names, such as the Mare Tranquillitatis (Sea of Tranquillity), Mare Nubium (Sea of Clouds), Mare Crisium (Sea of Crises), Oceanus Procellarum (Ocean of Storms), Sinus Iridum (Bay of Rainbows), Lacus Somniorum (the Lake of the Dreamers) and so on. Originally it was thought that they were true seas, or at least sea-beds. Studies of samples brought home from the Moon have shown otherwise; there has never been any water there, and suggestions that there may be sub-crustal deposits of ice are singularly unconvincing.

The Moon's lack of atmosphere makes it a hostile place, but nothing else was to be expected. The escape velocity is only 1½ miles per second, so that any sort of conventional atmosphere would leak away quite quickly. Without air, there can be no water. The lunar seas are dry lava-plains without a trace of moisture in them.

Many of the larger maria are basically circular in outline, and most of them make up a connected system which does not spread on to the far side; the Mare Crisium is the main exception. The vast, well-marked Mare Imbrium or Sea of Showers is easily visible with the naked eye, and has an area equal to that of Great Britain and France combined. The walls of the circular maria are raised into mountain chains such as the Apennines, Alps and Caucasus; the Apennines, much the most spectacular of the ranges, form part of the border of the Mare Imbrium, and include peaks which tower to 15,000 feet above the land below. Isolated peaks abound; no part of the Moon is free from them.

The entire lunar scene is dominated by the craters, which range from huge enclosures well over 100 miles in diameter down to tiny pits too small to be seen from Earth. Copernicus, a 56-mile crater near the edge of the Mare Nubium, has massive, terraced ramparts rising to over 12,000 feet above a sunken floor upon which there is a group of central mountains. Not far away from it is Stadius, of comparable size, but low-walled and hard to trace; it gives every sign of having been overwhelmed by lava. Fracastorius, on the border of the Mare Nectaris (Sea of Nectar) is 75 miles across, but has lost its seaward wall, so that it has been transformed into a huge bay; another bay is Hippalus, bordering the Mare Humorum (Sea of Humours) which still shows the remnant of a central peak. There are craters with relatively flat floors; craters with bright interiors, others with floors so dark that under suitable lighting conditions they look like pools of ink; craters which are regular and well-defined, and others which have been so broken and deformed that they are scarcely recognizable. Craters frequently overlap each other, and are spread all over the Moon, from the grey seas to the bright uplands and even the slopes and summits of mountain peaks.

In general, the craters are named after personalities of the past, usually scientists – a convenient if rather controversial system introduced in 1651 by a Jesuit priest-astronomer named Riccioli, who drew a lunar map which was reasonably accurate by the standards of the time. Astronomers such as Ptolemy, Tycho, Kepler and Copernicus are all there (though Riccioli, who had no time for the absurd idea that the Earth revolves round the Sun, did admit that he had 'flung Copernicus into the Ocean of Storms'). Julius Cæsar is among those present, though he has been honoured for his calendar reform rather than his military prowess.

Among the more unexpected names are Birmingham, Billy and Hell, all of which commemorate past astronomers of considerable eminence. Needless to say, Riccioli's system has been extended since 1651 to accommodate later scientists, but the leading craters had already been used up, so that men such as Newton, Halley and Einstein have had to be content with second best and Galileo – no favourite of Riccioli's – has been fobbed off with a very obscure, broken-down crater not far from the edge of the Oceanus Procellarum.

Though some of the craters are deep, they are not in the least like steep-sided mine-shafts. A normal crater has a rampart which rises to a modest height above the outer surface, and a floor which is depressed well below the mean level. Where there is a central peak, we find that the summit never rises as high as the surrounding walls (as has been correctly said, one could put a lid over it). Also, lunar slopes are surprisingly gentle, and because the Moon is so much smaller than the Earth its surface curves more sharply.

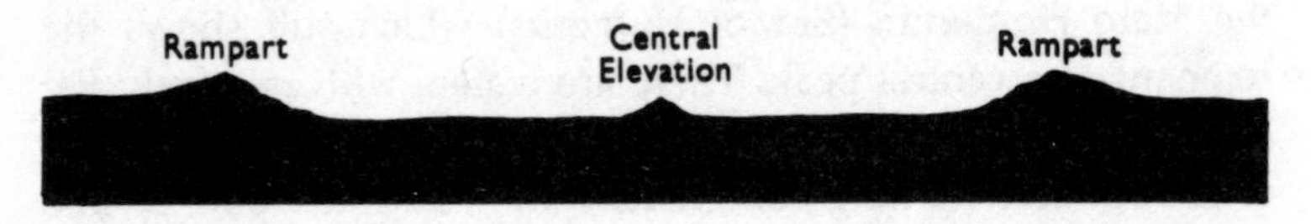

Cross-section of a large crater. The central peak never attains the height of the outer ramparts.

The depths of the craters – and, for that matter, the heights of the mountains – are measured by the lengths of the shadows they cast. When the Sun is low over a crater, the interior will be filled with shadow, making the crater itself very conspicuous. The effect is particularly well seen with the Sinus Iridum or Bay of Rainbows, which leads out of the Mare Imbrium. If it is observed at the right moment, the mountainous rampart will seem to project out of the darkness, producing an appearance which has been nicknamed 'the jewelled handle'.

It is fascinating to follow the progress of sunrise over different parts of the Moon. A crater which may be very prominent one

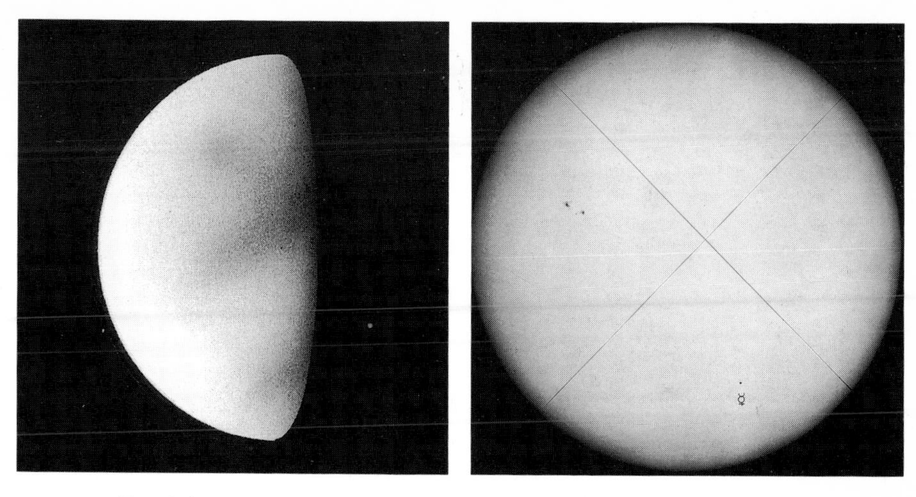

Above Left: Mercury from Earth, 3 June 1980, (Paul Doherty, 15-in refl. × 300).

Above Right: Transit of Mercury, 7 November 1914. (Royal Greenwich Observatory, 4-in photoheliograph. Reproduced by kind permission of the Royal Astronomical Society.) Note that Mercury (☿) is much darker than the sunspots.

Above Left: Mercury from Mariner 10. This is a mosaic from Mariner images.

Above Right: The ray-crater Kuiper on Mercury: Mariner 10. Kuiper is to the centre of the picture.

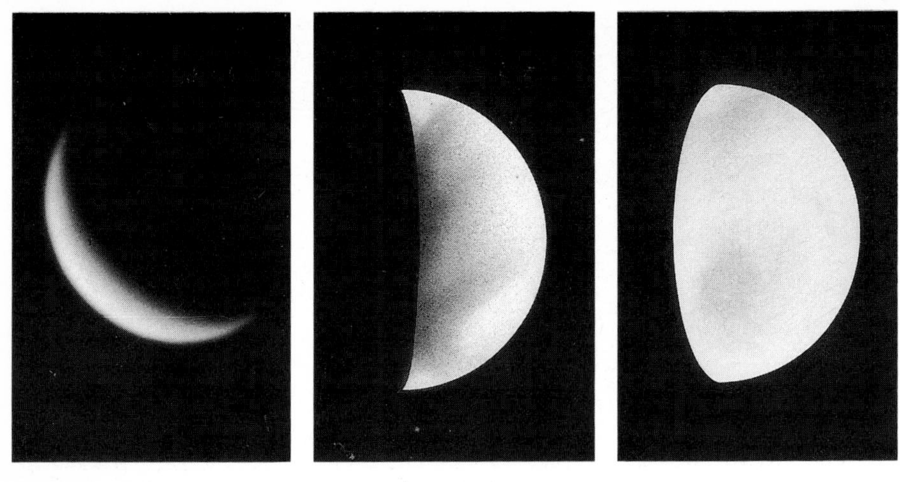

Above Left: Photograph of Venus, 15 March 1961, 15h. (H. E. Dall, 15-in refl.).

Above Middle: Drawing of Venus, 30 May 1985, 14.50 (Paul Doherty, 5-in OG × 250).

Above Right: Drawing of Venus, 25 December 1991 (Patrick Moore, 15-in refl. × 250).

Venus from Venera 10. Part of the grounded spacecraft is shown.

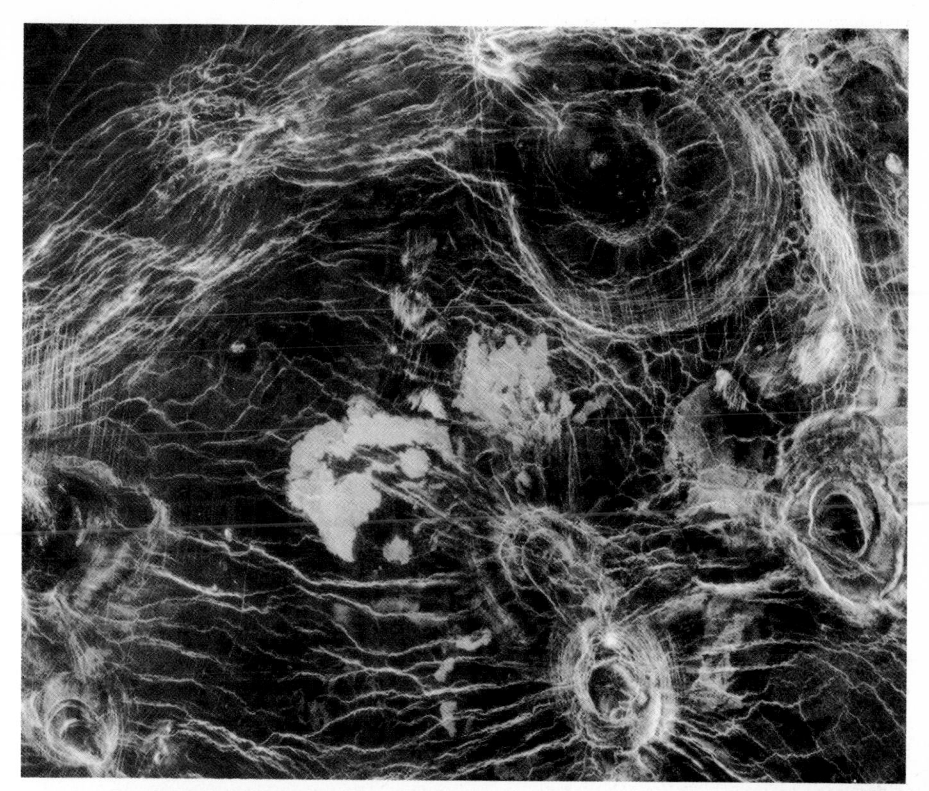

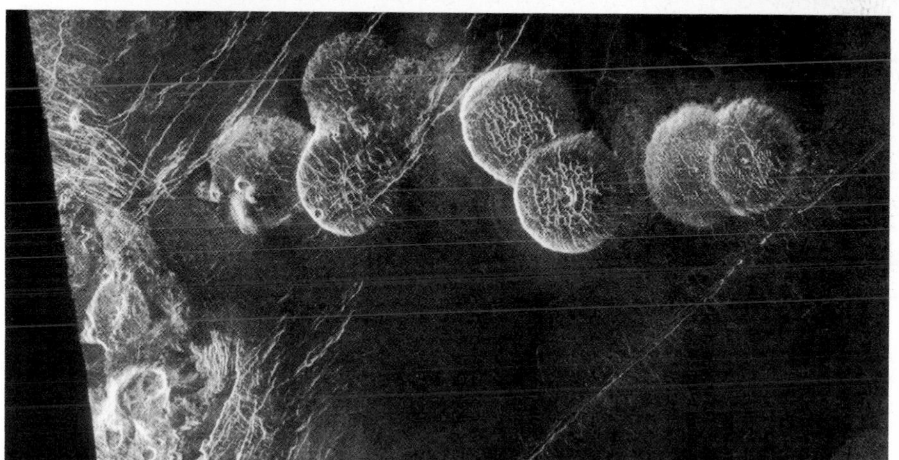

VENUS FROM MAGELLAN

Arachnoids on Venus! These range from 50 to 150 miles in diameter, and are circular volcanic structures surrounded by ridges, grooves and radial lines. The bright patches in the centre of the image are lava flows.

Eastern edge of Alpha Regio. Seven circular, dome-like hills around 15 miles in diameter are shown; they are probably thick lava-flows which came from an opening on relatively level ground, so that the lava flowed from the opening in an even pattern.

Far Top: Meteor Crater, Arizona, from its wall. This is a huge impact scar almost a mile wide.

Above: Aurora Borealis. These rayed bands are typical of a major display.

Above: Earthrise over the Moon. The Earth shows as a crescent in this picture, taken by the Apollo 14 astronauts while in lunar orbit.

Middle: The lunar Mare Nectaris, photographed by Patrick Moore (15-in refl.). The great bay of Fracastorius is shown at the top (south) of the Mare; to the right the great chain consisting of Theophilus, Cyrillus and Catharina.

Bottom: The Mare Imbrium, photographed by Commander H. R. Hatfield (12-in refl.). Plato to the lower left; Archimedes to the upper left. The Sinus Iridum (Bay of Rainbows) leads off from the Mare Imbrium to the bottom (north) of the picture. A few rays from Copernicus are shown near the top.

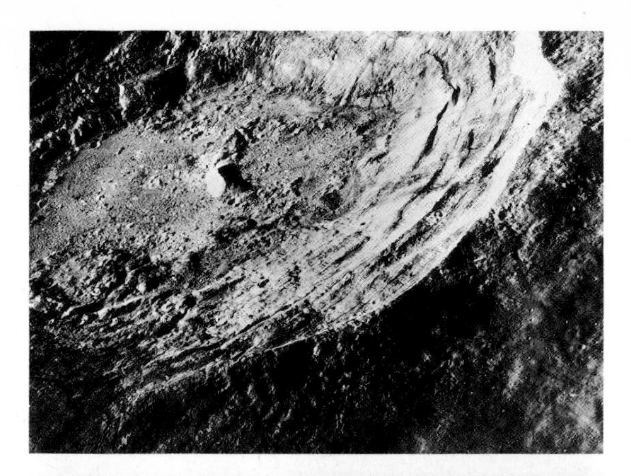

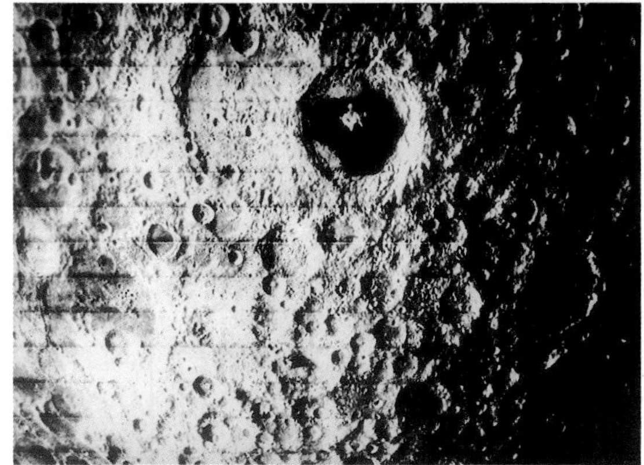

Top: The crater Aristarchus, from Apollo 15. The central elevations and the terraced walls are well shown. Though only 23 miles in diameter, Aristarchus is the brightest crater on the Moon.

Middle: The far side of the Moon, from Orbiter 3. The huge dark-floored formation is Tsiolkovskii; the central peak rises from a solidified lava-lake.

Bottom: A classic picture: Colonel Edwin Aldrin with the American flag on the Moon, 21 July 1969. The flag is motionless; there is no air to make it flutter!

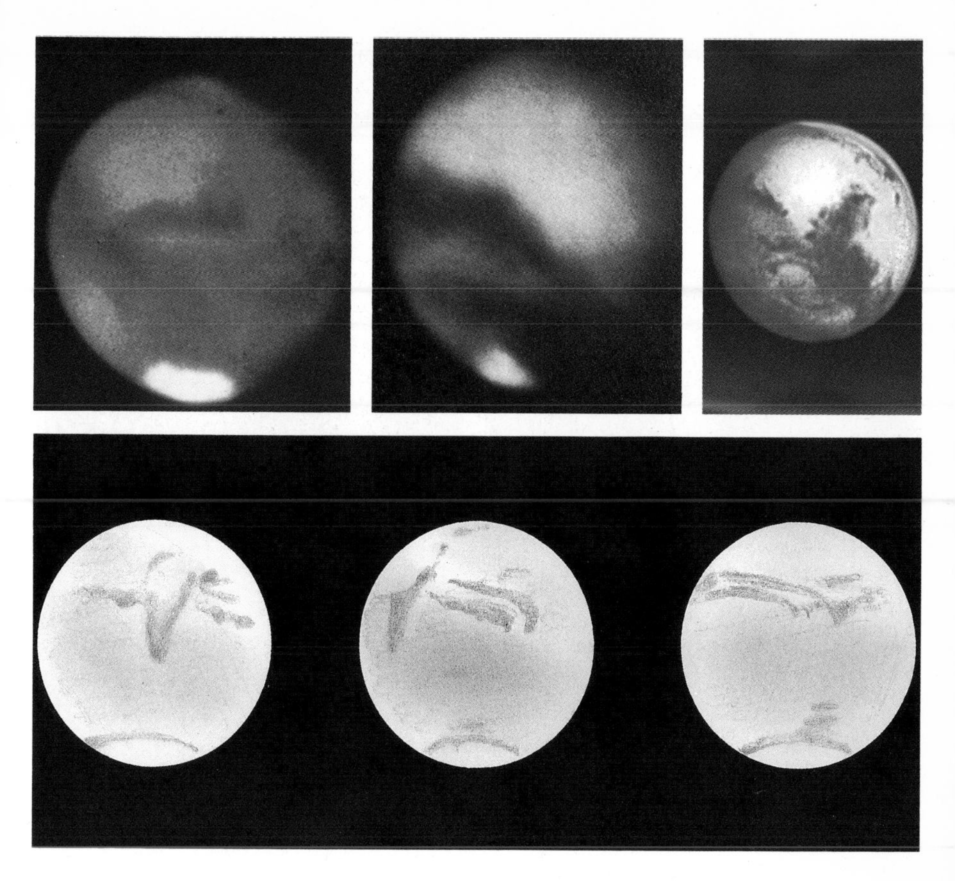

MARS FROM EARTH

Top Left: Photograph by C. F. Capen (24-in OG, Lowell Observatory). The south polar cap is to the top; Syrtis Major to the extreme left; Sinus Meridiani at the centre.

Top Middle: Photograph by W. S. Finsen (27-in OG, Johannesburg), showing the south polar cap and the Mare Sirenum area.

Top Right: Photograph from the Hubble Space Telescope. Syrtis Major is dominant.

Above: The rotation of Mars; three drawings by Patrick Moore. *Left*: 2 January 1993, 21.10, Syrtis Major central. *Middle*: 3 January, 00.30; Syrtis Major to the right: 3 January, 0.00. Syrtis Major has now been carried over the limb (15-in refl. × 400).

Top Left: Mars from 348,000 miles; Viking 1, 17 June 1976. The great volcanoes of the Tharsis Ridge are seen to the left; Olympus Mons above them. North is to the upper right.

Top Right: Olympus Mons from Viking; the greatest volcano in the Solar System, topped by a complex caldera. The width of the base is 300 miles.

Middle: Valles Marinens (Mariner Valley), from Viking. This is the largest of the Martian canyons.

Bottom: The scene from Viking 2 Lander, in Utopia. Large blocks litter the surface; on the right horizon, flat-topped hills are lit by the afternoon sun.

night, when it is on the terminator and is shadow-filled, may become very obscure later, when the shadows inside it have almost disappeared. At full moon, when the shadows are at their shortest, even large, deep craters become hard to identify unless they are unusually bright-walled or dark-floored. Full moon is the very worst time to start trying to find one's way about, and the most spectacular views are to be had between the crescent and three-quarter phase.

Of course, there are some special features which are easy to recognize at any time. Such is Plato, a 60-mile crater with a dark floor; Hevelius, a mid-seventeenth century lunar observer, called it 'the Greater Black Lake'. Near the Moon's limb there are two more very dark-floored formations, Grimaldi and Riccioli.* All these are basically circular, but since they lie well away from the apparent centre of the Moon they are foreshortened into ellipses. Very close to the limb, foreshortening becomes so marked that it is difficult to tell a crater from a ridge, and before the age of orbiting space-probes the limb regions were poorly mapped.

There are also some really bright craters. The most brilliant of all is Aristarchus, which is always conspicuous even though it is a mere 23 miles in diameter; it can often be seen when it is on the dark side of the Moon, and lit up only by earthlight. Proclus, near the well-marked Mare Crisium, is another very bright crater. But near full phase, most lunar features are overpowered by the bright streaks or rays which issue from a few craters, notably Tycho in the southern highlands and Copernicus, near the junction of the Mare Nubium and the Oceanus Procellarum.

The rays cast no shadows; they are surface deposits, and they are well seen only under high light. When near the terminator, Tycho looks like a normal bright-walled crater, 54 miles in diameter, with terraced walls and a central peak. Near full, it and Copernicus are the two most striking objects on the entire surface. There are numerous minor ray-centres, many of which are 'hot spots' cooling down less rapidly than their surroundings during the lunar night and during times of eclipse by the shadow of the Earth.

* No prizes are offered for guessing who named this particular crater, which is larger than Wales. Its even bigger neighbour is named Grimaldi, after Riccioli's pupil.

Other lunar features include the valleys, such as the great gash which slices through the Alps on the border of the Mare Imbrium; the rills or clefts, which are collapse features and look superficially like the cracks in dried mud; the chains of small craterlets which have been nicknamed 'strings of beads', and the gentle swellings known as domes, many of which have summit pits. Of special note is the Straight Wall in the Mare Nubium, which is not straight and is not a wall; it is a fault, 80 miles long, which casts a black line of shadow before full moon and appears as a bright line afterwards, when the Sun's rays fall on its face. Anyone who equips himself with a telescope and an outline lunar map will soon learn how to recognize the various features, and there is always something new to see.

It is amazing now to look back and see how little we really knew about the Moon before the rockets started to fly. There was even one strange theory according to which the 'seas' were covered with soft dust drifts several miles deep, so that any spacecraft unwise enough to land there would promptly 'sink out of sight with all its gear', to quote one eminent astronomer. All this was changed in a few years. The Russians led the way with their Luniks of 1959; then came the American Rangers, which crash-landed on the Moon and sent back close-range pictures during the last few minutes of their flight. There were the five highly successful Orbiters, which went round and round the Moon and sent back superb pictures of both the near and the far sides; and there were the soft-landers, of which Russia's Luna 9 was the first. But everything paled before the triumph of 1969, when *Eagle*, the lunar module of Apollo 11, touched down on the lunar surface carrying two astronauts, Neil Armstrong and Edwin Aldrin. Other Apollo missions followed, and the programme ended in December 1972, when Eugene Cernan, commander of Apollo 17, became the last man to leave the Moon. Since then there have been only a few minor Russian unmanned probes, plus one Japanese orbiter (Hagomoro) which was really in the nature of a first attempt. This is not the place to go into details about the Apollo missions, so I will merely attempt to give a very brief summary of the main results.

First, the Moon really is airless; the atmosphere – if we can call it that – is so tenuous that it can be totally ignored. There is an

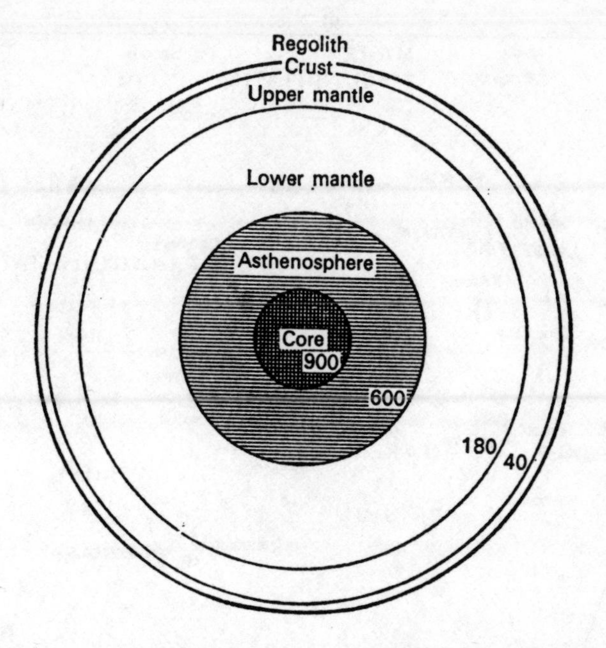

Internal composition of the Moon; core, asthenosphere (region of partial melting), mantle, crust and regolith. The figures indicate the depths, in miles, below the visible surface.

upper loose surface layer or regolith, on average 13 to 16 feet deep below the maria and 30 feet below the highlands. Underneath the regolith is a half-mile thick layer of shattered bedrock, and then comes a layer of more solid rock down to a depth of around 16 miles. Various other layers follow, and finally we come to the metal-rich core, which may be from 600 to just over 900 miles in diameter, and is hot enough to be molten.

All the rocks brought home for analysis are volcanic; the average age of the highland rocks is from 4 to 4.2 thousand million years. Mild moonquakes are common, most of them deep-centred although some are shallow. The lack of an overall magnetic field was confirmed, though there are areas of locally magnetized material, and it is quite likely that the Moon did once have a general magnetic field which has now died away. Beneath some of the circular maria and also below some of the largest craters there are high-density areas known as mascons (from *mass concentrations*), detected by irregularities in the movements of orbiting

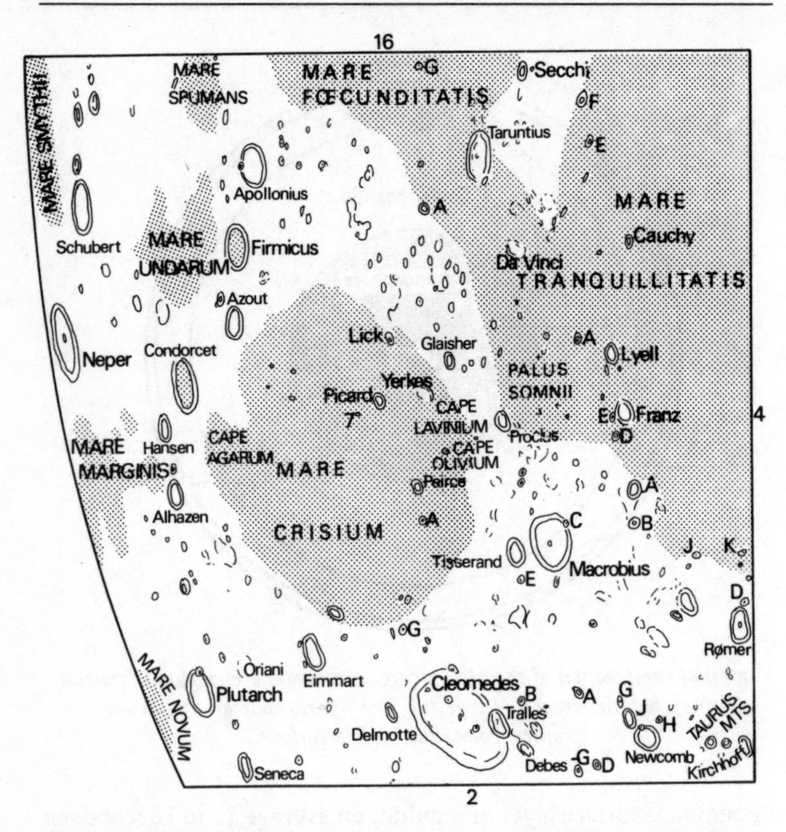

Section of my 2-ft map of the Moon. It shows the Mare Crisium (Sea of Crises) which is separate from the main Mare system, as well as parts of the Mare Tranquillitatis (Sea of Tranquillity) and the Mare Fœcunditatis (Sea of Fertility). Smaller seas are the Mare Undarum (Sea of Waves), Mare Spumans (Foaming Sea), Mare Smythii (Smyth's Sea) and Mare Novum (New Sea). Craters include the large walled plain Cleomedes and the brilliant ray-crater Proclus.

lunar spacecraft. Presumably these are condensations of volcanic rock.

There were some particularly intriguing episodes. One of these occurred during the Apollo 17 mission, when one of the two 'moon-walkers' was Dr Harrison Schmitt, a professional geologist who had been trained as an astronaut specially for the trip. Suddenly he announced the discovery of 'orange soil'. The effect at Mission Control in Pasadena (where I was) was electrifying;

could we have found indications of current volcanic activity? Not so; the orange hue was later proved to be due to nothing more exciting than numbers of small, very ancient coloured glassy droplets.

It was also confirmed that there is no trace of hydrated material, so that the Moon has always been bone dry. It seems strange to recall that as recently as 1966, at an international meeting in Prague, one of the world's greatest planetary geologists, Professor Harold Urey, was trying (unsuccessfully) to convince me that the lunar seas had once been oceans filled wth ordinary water.

This brings me on to the question of how the various lunar

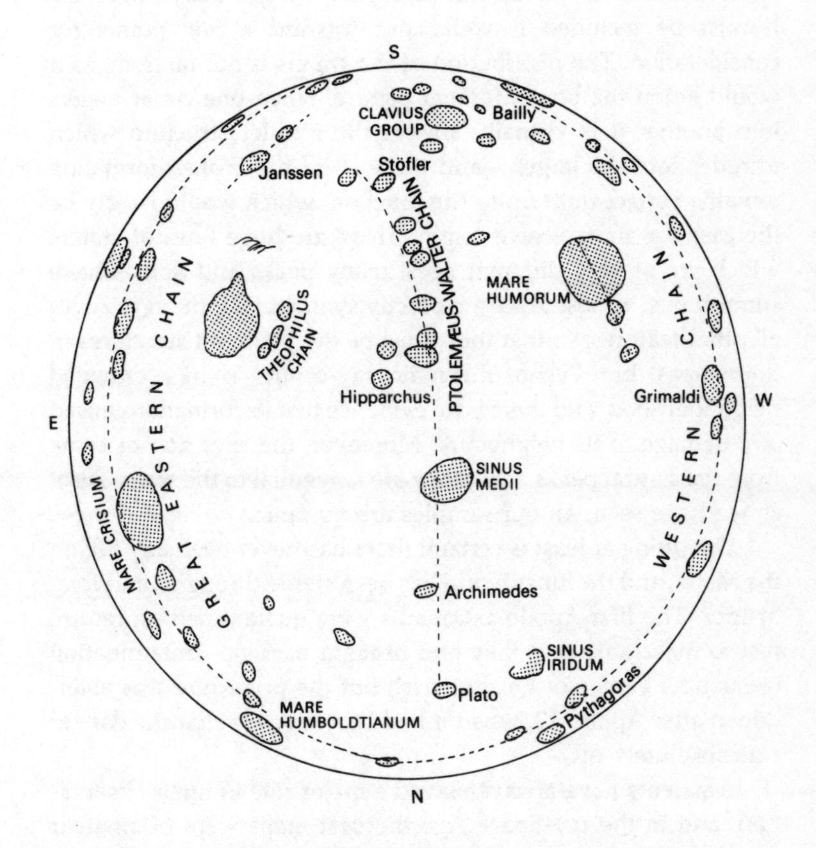

Non-random distribution of the large lunar features – though they are of very different ages.

features were formed. For many years there have been fierce arguments between those who believe that the maria and the craters are due to impact, and those who maintain that they are of internal origin, so that they are related to terrestrial formations of the caldera type. Today it is the impact theory which is favoured by almost all astronomers. The Moon suffered a violent cosmical bombardment, which ended with the creation of the youngest seas around 3.8 thousand million years ago. There followed an upsurge of lava from below the crust, which flooded the basins. Since then very little has happened apart from the occasional formation of a new impact crater; even the youngest features of the Moon are very ancient on the terrestrial time-scale.

Supporters of the volcanic theory are few nowadays, but since I must be included I would put forward a few points for consideration. The distribution of the craters is not random, as it would be on the bombardment picture; when one crater breaks into another it is virtually always the smaller structure which intrudes into the larger – and the wall of the broken formation remains perfect right up to the junction, which would hardly be the case for an explosive origin. There are huge lines of craters which are of very different ages; many peaks and domes have summit pits, almost always perfectly symmetrical; the rays cover all other features, so that they must be the youngest structures in their areas, but Tycho, the main ray-centre, is in a crowded neighbourhood, and there is no evidence that its formation caused any damage to its neighbours. Moreover, the rays do not come from the central peaks at all; they are tangential to the walls. Also, as we have seen, all our samples are volcanic.

One thing at least is certain: there has never been any life on the Moon, and the lunar world has been sterile throughout its long history. The first Apollo astronauts were quarantined on return, just to make sure that they had brought back no contamination (remember Professor Quatermass!) but the procedure was abandoned after Apollo 12, when it had become clear that the danger was absolutely nil.

Amateurs have always played a major rôle in lunar observation, and in the pre-Space Age the best maps were of amateur construction. It was also possible to make interesting discoveries, particularly in the libration regions. After Apollo there was a general feeling that observations from Earth could add little to what the astronauts had told us, but this is not entirely true, and in particular there are the so-called TLP or Transient Lunar Phenomena.

TLP (a term for which I believe I was responsible) take the form of localized, usually short-lived glows or obscurations. They are most often seen in regions rich in rills, or around the boundaries of the circular maria; of special note is the brilliant crater Aristarchus, while many obscurations have been reported inside the dark-floored Plato. Searching for them is time-consuming and laborious, but it is well worth while. Probably they are due to the release of gases from below the crust, but our information is still very scanty, and only one good spectrum has been obtained, by the Russian astronomer N. A. Kozyrev in 1958, when he saw a red glow inside the prominent walled plain Alphonsus. If we are to learn more about them, amateurs must take the lead. It would be important to give final proof that dormant though it may be, the Moon is not totally inert, even though we can be sure that there have been no major structural alterations for at least a thousand million years and probably longer.

Nowadays there is nothing far-fetched in the idea of a full-scale base on the Moon. As a laboratory, an observing station and in many other ways it would be of tremendous benefit to mankind; and if all goes well, it will happen. I will always remember the words of Neil Armstrong, when he joined me for a television programme some years ago:

'I'm quite certain that we'll have such bases in our lifetime. Somewhat like the Antarctic stations and similar scientific outposts, continually manned . . . and in some ways the Moon is more hospitable than the Antarctic. There are no storms, no snow, no high winds, no unpredictable weather; as for the gravity – well, the Moon's a very pleasant place to work in; better than the Earth, I think.'

Look at the Moon now, on any clear night, and you can see the grey maria, the mountains, the valleys and the craters shining down from a quarter of a million miles away. It is sometimes hard to credit that men have actually been there. Yet with all this, I hope you will agree that the Moon has lost none of its magic.

CHAPTER NINE

Mars

Beyond the Earth–Moon system we come to the red planet Mars, which has always had a special interest for us. Even in our own century it was thought quite possible that there might be intelligent life there, and there were various stories about 'signals from the Martians' which were taken quite seriously. When this idea had to be given up, it was still thought that Mars could support a great deal of vegetation. This was certainly my own view; before 1965 I would have rated the chances of Martian life at well over eighty per cent.

We know better now. Since the first unmanned probe by-passed Mars, our whole picture of the planet has had to be changed. We still cannot be absolutely certain that Mars is totally sterile, but the evidence does point that way. First, a few facts and figures may be useful.

Mars moves round the Sun at a mean distance of 141,500,000 miles. Its orbit is decidedly eccentric, and the actual distance ranges between 154,000,000 miles at aphelion down to only 129,500,000 miles at perihelion. This has a marked effect on the seasons during the 687-day long Martian year. Because the axial tilt is much the same as ours (24 degrees, as against 23½ degrees for the Earth), southern summer there falls when Mars is at its closest to the Sun, so that summers in the southern hemisphere are shorter and hotter than those in the north, while the winters are longer and colder. As is only to be expected, Mars is a chilly world. On a 'hot' summer day at the equator the temperature may rise to 10 degrees Centigrade (50 degrees Fahrenheit), but any Martian night is much colder than a polar night on Earth, and long before sunset a thermometer will drop far below freezing-point. The axial rotation period is 24 hours 37 minutes 22.6 seconds – a value which is known very accurately, because the surface markings are permanent, and we can watch Mars spin. Since Viking times, when the first soft-landing probes came down on the Martian surface, a 'day' there has become generally known as a 'sol'.

The calendar is straightforward enough – unlike that of Venus! – and I think it is best summarized in a table:

MARTIAN SEASON	LENGTH EARTH DAYS	LENGTH SOLS
Southern spring (northern autumn)	146	142
Southern summer (northern winter)	160	156
Southern autumn (northern spring)	199	194
Southern winter (northern summer)	182	177
	687	669

(Working out a Martian calendar is a pleasant pastime, even if somewhat premature as yet. My own suggestion is to divide up a Martian year into 18 months, each of 37 sols. This makes 666 sols. We need 669, so an extra sol can be tacked on to Months 6, 12 and 18, giving them 38 sols instead of 37. This should work reasonably well, though there will have to be an adjustment to allow for the fact that the revolution period of Mars is not exactly 687 days, but 686 days 23 hours 52 minutes 31 seconds. Naming our eighteen months is not something to be considered yet, though no doubt there will be the usual tedious political squabbles when it eventually has to be tackled!)

Mars comes to opposition at intervals of around 780 days. Thus there are oppositions in January 1993, February 1995, March 1997 and April 1999. The orbital eccentricity means that not all oppositions are equally favourable. In 1988, for example, opposition fell when Mars was near perihelion, and the minimum distance from Earth was only 36,300,000 miles; in 1995 opposition occurs near Martian aphelion, and the least distance from Earth will be a full 63,000,000 miles. At its best, as in 1988, Mars can outshine every other natural body in the sky apart from the Sun, the Moon and Venus, but when well away from opposition it falls to around the second magnitude, comparable with the Pole Star.

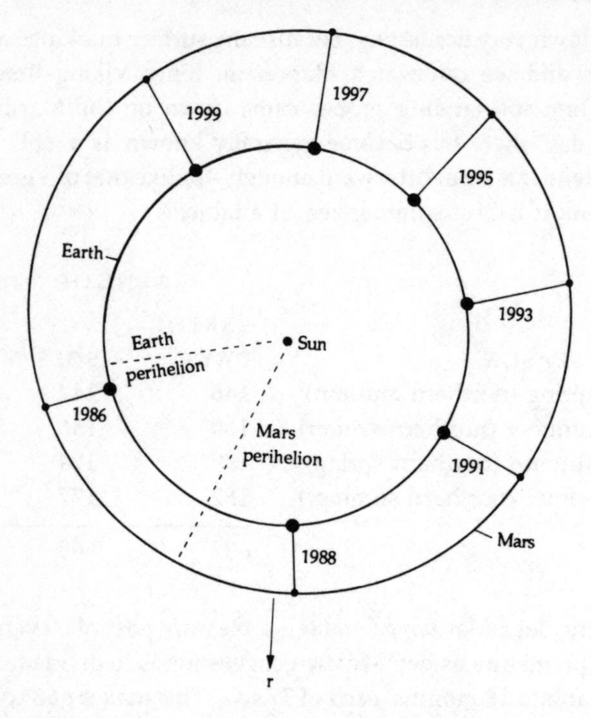

Oppositions of Mars, 1986–1999. Obviously the 1986 and 1988 oppositions are very favourable, while those of 1995 and 1997 are unfavourable.

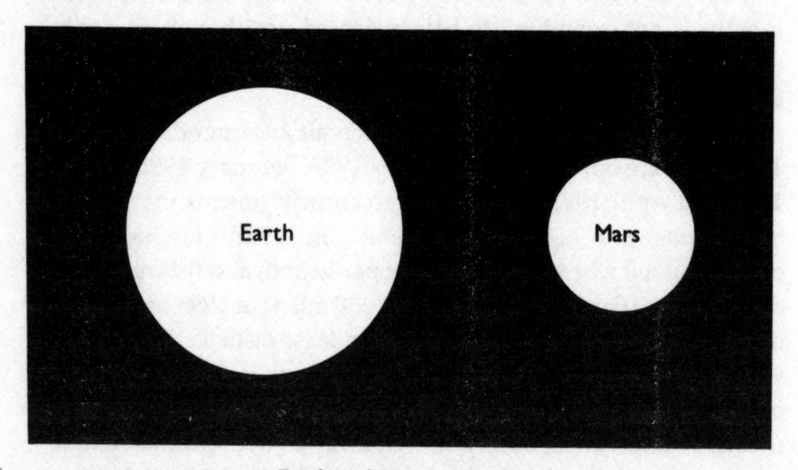

Earth and Mars compared.

(In 1719 it is recorded that Mars was particularly brilliant, and caused alarm among people who mistook it for a red comet which was about to hit the Earth!)

Telescopically, Mars can show an appreciable phase, and can appear the shape of the Moon a few days before or after full. For obvious reasons it never becomes a half or a crescent – at least, not as seen from Earth.

Although it is so close to us on the astronomical scale, Mars is not so easy to observe as might be thought. It is a small world, with a diameter of only 4222 miles, so that in size it is intermediate between the Earth and the Moon. Except when it is near opposition, only large telescopes can show much detail on its surface, and it is understandable that in the pre-Space Age we were forced into a good deal of speculation.

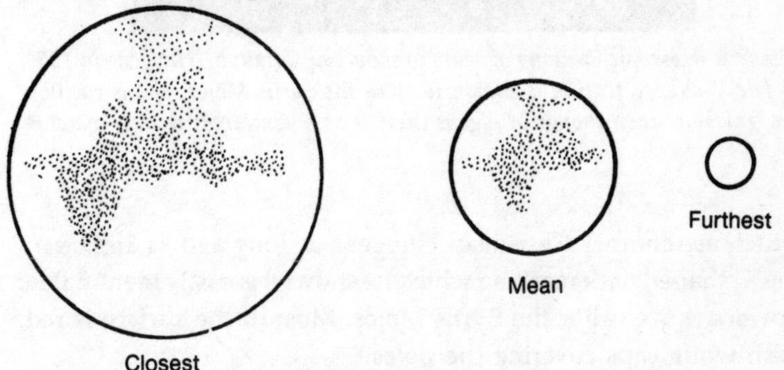

Changing apparent size of Mars; it ranges between over 25 seconds of arc and less than 4 seconds of arc.

The Earth, with its relatively large mass and high escape velocity, can hold on to a dense atmosphere; the Moon has none. Mars, then, would be expected to have a thin atmosphere, and this is exactly what is found, though we now know it to be much more tenuous than we believed before 1965. There has never been any serious suggestion that Earth-type creatures such as ourselves could breathe there, and the 'Martians' so popular with science-fiction writers were always assumed to be built upon a different and unfamiliar pattern.

Mars differs from Venus inasmuch as its visible markings are clear-cut and basically permanent. They were first drawn by the

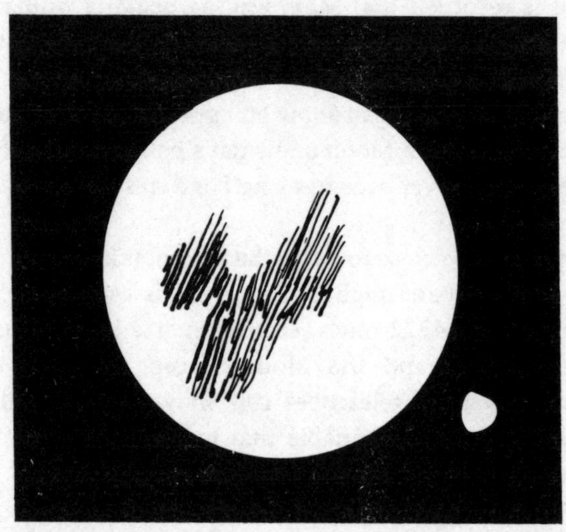

The first telescopic drawing of Mars to show any detail: by Huygens in 1659. The V-shaped feature is known to us as the Syrtis Major, and is easily recognizable, even though Huygens drew it as much larger than it really is.

Dutch astronomer Christiaan Huygens as long ago as 1659, and the V-shaped dark marking which he showed is easily identifiable; nowadays we call it the Syrtis Major. Most of the surface is red, with white caps covering the poles.

The first maps of Mars were drawn in the first half of the nineteenth century. By around 1870 the maps had become reasonably good, and the features had been given names which were decidedly attractive. In general the dark areas were regarded as seas, while the red regions were land; the names honoured observers of the planet – thus we had Mädler Continent, Lassell Land, Lockyer Land, Beer Continent and so on. (This latter name commemorated Wilhelm Beer, one of the pioneer lunar and planetary observers.) Then, in 1877, the Italian astronomer Giovanni Virginio Schiaparelli, using a fine 8.6-inch refractor under the clear skies of Milan, made a new series of observations and revised the nomenclature; out went Beer Continent, Lockyer Land and the rest, and in came Solis Lacus, Chryse, Utopia and Margaritifer Sinus, as well as the triangular Syrtis Major.

By that time it had become clear that the Martian atmosphere

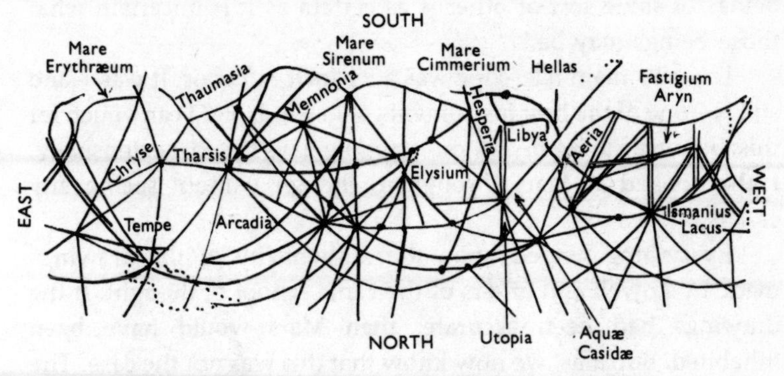

The canal network, according to Lowell. We know that the canals were simply optical illusions.

is too thin and too dry to allow oceans to exist, and it was more generally believed that the dark areas were old sea-beds filled with vegetation, though swamps were also suggested. Schiaparelli drew them carefully, but also drew attention to features which were much harder to explain. Crossing the ochre-red 'deserts' he showed straight lines which he called *canali*, or channels; inevitably this was translated into English as canals, and the myth of the Martian canals was born. Certainly Schiaparelli's maps looked very strange. The canal network seemed to be almost geometrical in aspect, and things were made even more curious when Schiaparelli announced that some of the canals had become double, so that a single streak was replaced by perfect twins.

For some time nobody else could see the canals, but they were confirmed in 1886 by two French observers, Perrotin and Thollon, using the powerful telescope at Nice. Subsequently, canals became all the rage. Schiaparelli himself kept an open mind about their origin, but this was not true of Percival Lowell, a wealthy American who set up an observatory at Flagstaff, in Arizona, specially to study the planet. Between 1895 and his death in 1916 Lowell made hundreds of drawings, showing a canal system which could not be natural. He was utterly convinced that it was an artificial irrigation network, built by the inhabitants to carry water through from the icy poles to the dry regions closer to the equator. He went so far as to write: 'That Mars is inhabited by beings of some sort or other is as certain as it is uncertain what those beings may be.'

Lowell's main telescope was a 24-inch refractor. It was – and still is – one of the best instruments of its kind, and I can vouch for this, because during my Moon-mapping days I used it extensively. I also looked at Mars through it, though without seeing any canals . . .

Everything hinged upon the authenticity of the drawings made by Lowell and others of the canal school of thought. If the drawings had been accurate, then Mars would have been inhabited. But, alas, we now know that this was not the case. The canals could never be properly photographed; all that could be made out on the plates were ill-defined features which did not look in the least artificial. The problem was finally solved when the first spacecraft flew past Mars in 1965, and sent back pictures from close range. Canals were conspicuous only by their absence. They were tricks of the eye – not really surprising when one is trying to glimpse detail at the very limit of visibility. It is also true that Lowell, great man though he was in many respects, was not a reliable observer; he drew streaky features not only on Mars but also on Mercury, Venus and the satellites of Jupiter.

There is a lesson to be learned here. Once the canal network had been publicized, many observers saw it – or thought they did. I have even seen canals shown on sketches made with telescopes as small as 6 inches aperture. It is only too easy to 'see' what one half-expects to see.

I carried out an investigation of my own some years ago. Initially I thought that the canal illusion might be due to mountain ridges, chains of craters, valleys, or even the borders between areas of different brightness, so I took a modern map, of unquestioned accuracy – based on the spacecraft results – and upon it superimposed Lowell's canal network. There was no correlation whatsoever.

On the other hand, there is considerable real detail to be seen whenever Mars is well placed. The problem is that a high magnification is needed, so that the average amateur-owned instrument can be properly used only for a few months to either side of opposition.

Before 1965, the year of the first successful Mars flight, it was

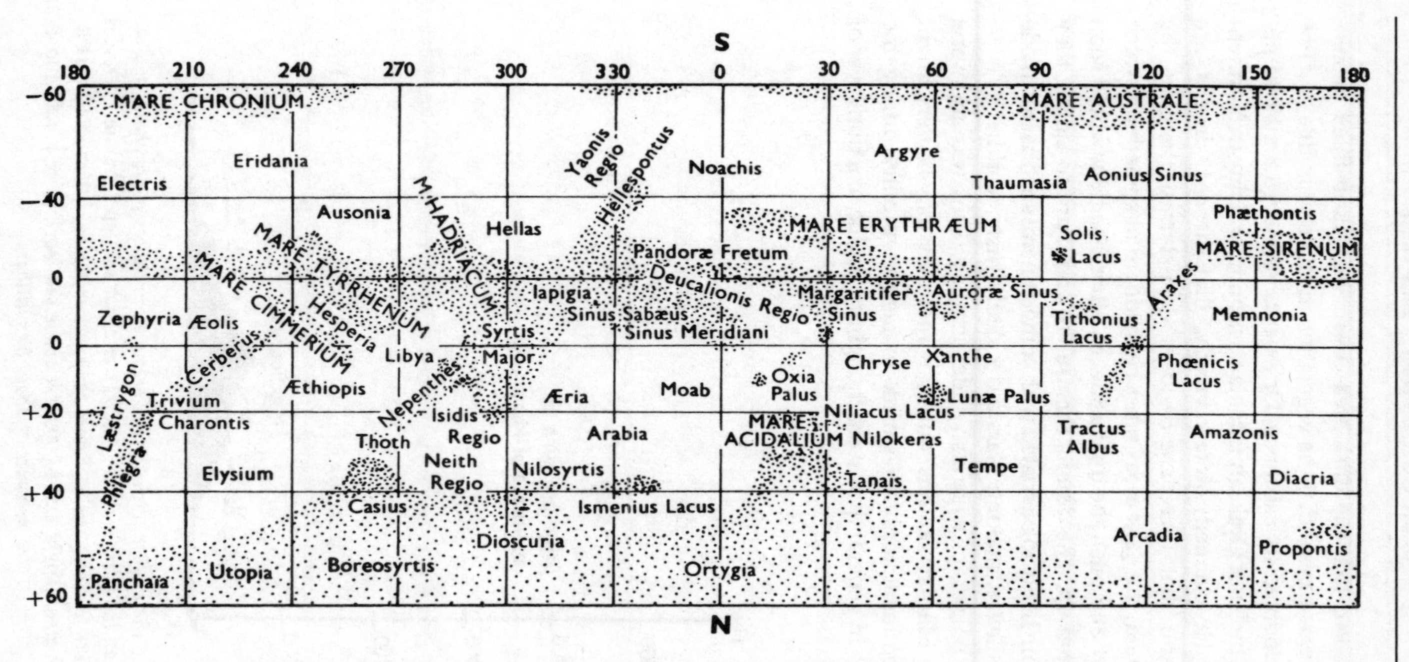

Map of Mars: Mercator projection. This has been drawn from my own observations with 8.5-in, 12.5-in and 15-in reflectors. I have followed the classic nomenclature, which was based on Antoniadi's.

still believed that the dark areas were vegetation-filled depressions; it was agreed that the vegetation must be primitive, but few people doubted its existence. One persuasive argument was put forward by Ernst Öpik, an Estonian astronomer living in Northern Ireland. He claimed, quite correctly, that the red 'deserts' are dusty, and that there are frequent dust-storms; therefore, he maintained, the dark areas must be made up of something which can grow and push the dust aside, as otherwise they would soon be covered up. The deserts themselves could not be sandy; they were more likely to be coated with some mineral such as iron oxide of iron silicate, making Mars a decidedly rusty world.

The polar caps were also under scrutiny. They wax and wane with the Martian seasons; during winter they may be very bright and prominent, only to shrink to near-invisibility during the summer. In general they were thought to be due to a thin layer of

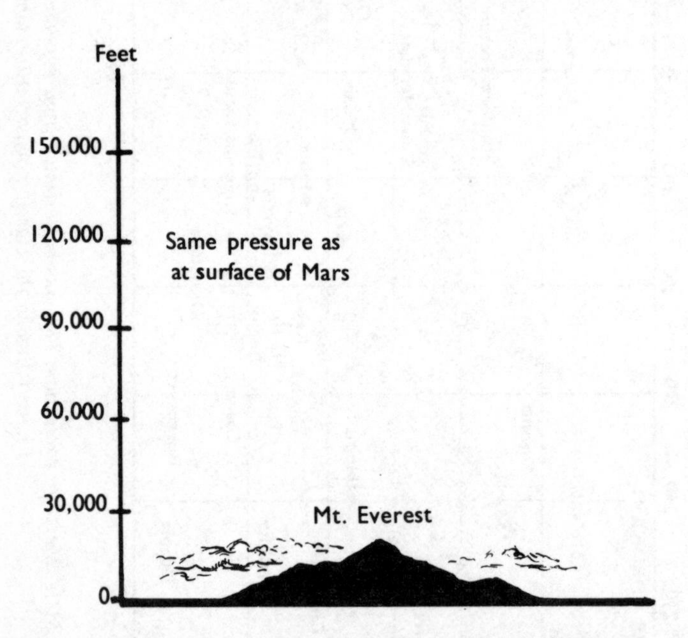

Density of the atmosphere of Mars: below 10 millibars everywhere, and no denser than the Earth's air at 120,000 feet above sea-level. Mount Everest is shown here for comparison. Before the space missions, it was thought that the equivalent height was 52,000 feet above sea-level, corresponding to a pressure of about 87 millibars.

hoar-frost, though there was also a theory that they might be caused by 'dry ice' – solid carbon dioxide. Finally, it was thought that the Martian atmosphere must be made up chiefly of nitrogen, with a ground pressure of around 87 millibars. This is equivalent to the pressure of the Earth's air at a height of rather less than twice that of the top of Mount Everest. There were no signs of mountains on Mars; the surface was expected to be rather gently undulating, with no major highlands anywhere.

Mariner 4 was launched from Cape Canaveral on 20 November, 1964. (Its twin and predecessor, Mariner 3, was a failure; it took off successfully, but contact was lost after a few hours, and was never regained.) Mariner 4 passed by Mars on 14 July 1965 at a range of 6000 miles, and in a few days it overturned many of our cherished ideas. The atmosphere was much thinner than had been expected; the ground pressure is below ten millibars everywhere, so that it corresponds to what we usually call a laboratory vacuum. The main constituent is not nitrogen, but carbon dioxide. The dark areas are not always depressions – the Syrtis Major is a high plateau – and they are not vegetation-covered; they are merely regions where the red dust has been blown away by Martian winds, exposing the darker material beneath. The polar caps really are ice, but are made up of a mixture of water ice and carbon dioxide ice; they are thick rather than being

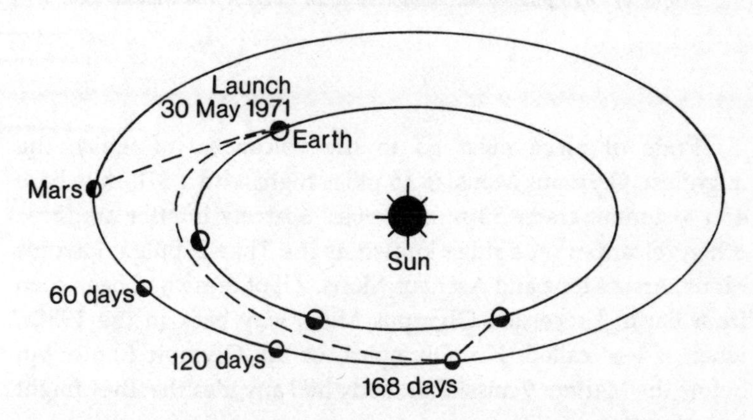

Flight path of Mariner 9. It was launched on 30 May 1971 and reached Mars 168 days later; it was then put into a closed path round Mars, and sent back pictures and data for almost a year.

mere surface layers. And, most significantly of all, Mars is a cratered planet. It seemed to be much more like the Moon than like the Earth.

Mariners 6 and 7 followed in 1969, confirming the first results; there were also some unsuccessful Russian vehicles (in fact, even today the Russians have had very little luck with Mars, and this is surprising in view of what they have accomplished with Venus, which should be a much more difficult target). Then, in 1971, Mariner 9 was put into a closed path around Mars, and operated for almost a year, sending back a series of magnificent pictures – 7329 in all. For the first time we had views of the Martian volcanoes, canyons, scarps, plains and channels.

Yet at first little could be seen, because Mars was experiencing one of its global dust-storms. Using a very powerful telescope (the 27-inch refractor at Johannesburg in South Africa) I had nights when I could see no surface detail at all. It was some weeks before the dust cleared away and we could see the Martian features in their full glory.

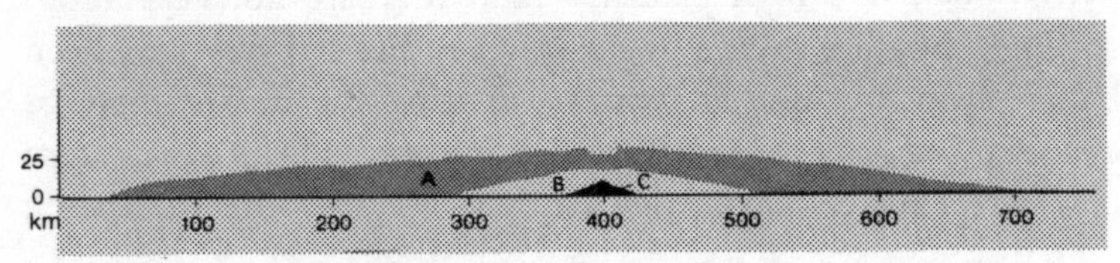

Profile of Olympus Mons, compared with Everest and Mauna Kea.

Pride of place must go to the volcanoes, of which the mightiest, Olympus Mons, is 15 miles high, with a 370-mile base and a summit crater 53 miles across. Scarcely inferior are three other volcanoes on a ridge known as the Tharsis Bulge: Pavonis Mons, Arsia Mons and Ascræus Mons. All of these had been seen from Earth; I recorded Olympus Mons way back in the 1930s, when it was called 'Nix Olympica' or the Olympic Snow, but before the Mariner 9 mission nobody had any idea that they might be volcanoes.

There are also canyon systems, such as the Valles Marineris, which can be traced for a total length of over 2800 miles, and is in places up to 370 miles wide and $4\frac{1}{2}$ miles deep; compared with this, our Grand Canyon of the Colorado pales into insignificance. Even more complex is the system of Noctis Labyrinthus, nicknamed the Chandelier, which will no doubt become a tourist attraction in the foreseeable future . . . The arrangement of canyons, when seen on a photograph, does indeed recall a chandelier – hence the nickname.

The two hemispheres of Mars are not alike. In general the southern part of the planet is the higher, more heavily cratered and more ancient, though it does contain two deep and well-formed basins, Hellas and Argyre. The northern hemisphere is lower, younger and less cratered, but it does contain part of the Tharsis Bulge. The Syrtis Major, the most prominent dark feature on the planet, lies just north of the equator; much further north is another very prominent dark area, the wedge-shaped Acidalia Planitia, formerly known as the Mare Acidalium. (Many of these names are due to Antoniadi. His map of Mars, unlike that of Mercury, was amazingly accurate.)

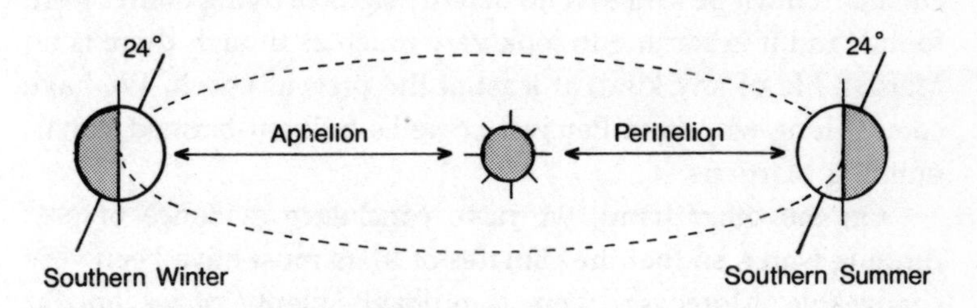

Martian seasons. The southern summer occurs near perihelion, and this makes a considerable difference, because the orbit of Mars is much more eccentric than that of the Earth.

Some of the features seem to be old river-beds; we can even see 'islands' in them. This means that in the remote past Mars must have been a friendlier and warmer place than it is now, with a thicker atmosphere and a good deal of running water. The polar caps are not identical; the southern cap is due mainly to carbon dioxide ice, covering a residual water-ice cap beneath, whereas the northern cap can lose its carbon dioxide coating in midsummer, leaving the lower cap visible. Remember that Mars' axial tilt is much the same as ours; and because southern summer falls near perihelion, the climate in the southern hemisphere is more extreme than that in the north.

The next step came in 1975, when two Viking spacecraft were sent to Mars; Viking 1 arrived in June 1976, and Viking 2 in the following August. Each probe was made up of an orbiter (to enter a closed path round the planet, serving both as a mapping vehicle and as a relay) and a lander (to come down gently, using parachutes and rocket braking). Both Vikings were successful. The first lander came down in the 'Golden Plain' of Chryse, north of the equator, while the second landed in the more northerly plain of Utopia. Both sent back spectacular pictures, showing red, rocky landscapes; the sky was pink, rather than dark blue as many people had expected, and the temperatures were very low, never higher than −13 degrees Centigrade (−24 degrees Fahrenheit) in Chryse and even chillier in Utopia. The winds were moderate. Material was scooped up from the desert, drawn in to the vehicles and analysed in a search for life; the results were not entirely clear-cut, but it must be said that no definite signs of living matter were found, and it is starting to look very much as though there is no Martian life of any kind, at least at the present epoch. We have come a long way from Percival Lowell's brilliant-brained, canal-building Martians.

On the other hand, we have conclusive evidence of past running water, so that the climates of Mars must have been very changeable. Moreover, there is probably plenty of ice not far below the crust, so that Mars, unlike the Moon, has not always been dry; we can also see traces of past flash-floods. In view of this, there must be a distinct chance that life did once exist there, and has now either died out or gone into prolonged hibernation. The only way to find out is to obtain samples from Mars and subject them to a very thorough analysis. I would not bet on the probability of finding any Martian fossils, but neither would I completely rule it out.

The latest probe, Mars Observer of 1993, is not a lander; its sole task is to improve our maps of the surface. We also have magnificent pictures of Mars from the Hubble Space Telescope, showing fine details. Bearing all this in mind, is there anything left for the Earth-based observer to do?

The answer is a resounding 'yes', because Mars is not a changeless, inert world. In particular, there are cloud phenomena. Isolated cloud systems are common, quite apart from the widespread dust-storms; we never know when or where to expect them, and it is important to track them, so that we can learn more about Martian meteorology. Variations in the outlines of the dark and bright areas can also be noted, together with the waxing and waning of the polar caps. Finally – and I admit that this is a long shot! – can we be absolutely certain that the Martian volcanoes are extinct? This cannot be proved; if there is a major eruption, then a well-equipped amateur will be probably the first to detect it. The chances may be slight, but they are are not nil.

Mars has two satellites, Phobos and Deimos, both discovered by Asaph Hall in 1877 after a prolonged search. Both are very

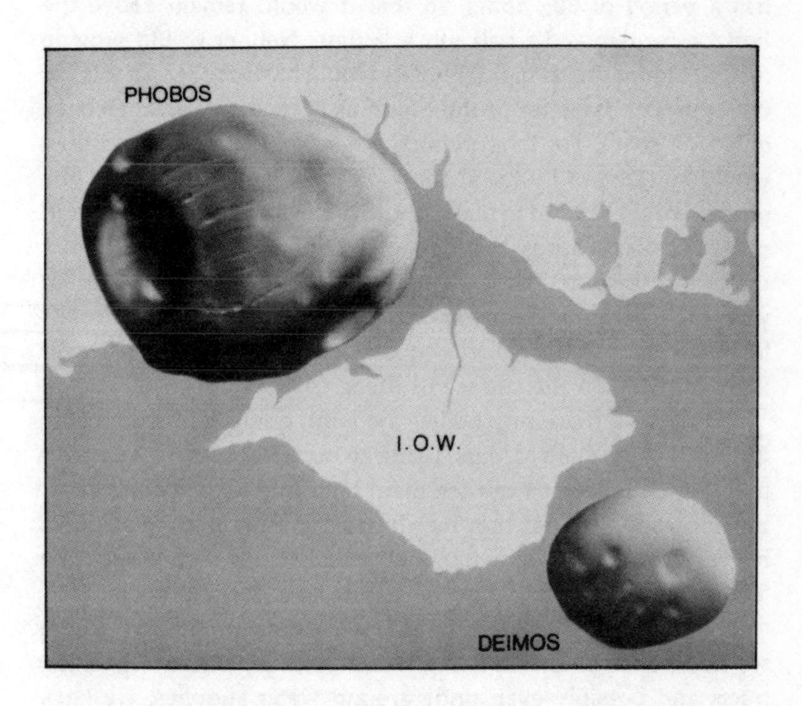

Sizes of Phobos and Deimos, compared with the Isle of Wight.

small, and are irregular in shape; Phobos measures 17 × 14 × 14 miles, Deimos only 6 × 7 × 9 miles. Both were studied from close range from Mariner 9 and the Viking orbiters, and both proved to have cratered surfaces; one crater on Phobos is as much as three miles in diameter, so that if it has been formed by the impact of a meteorite Phobos itself would have been in grave danger of being shattered. These dwarf attendants are quite unlike our massive Moon, and it seems very likely that they are ex-asteroids which have been captured by Mars.

Phobos moves at only about 3600 miles from the surface of Mars, and has a revolution period of 7 hours 39 minutes – less than a Martian sol; to an observer on the planet Phobos would rise in the west and set in the east 4½ hours later, during which time it would go through more than half its cycle of phases from new to full. The interval between successive risings would be just over eleven hours. Deimos, over 12,000 miles from the Martian surface, has a period of 30¼ hours, so that it would remain above the horizon for two and a half sols at a time. Neither would provide much illumination at night; Phobos would have about one-third of the apparent diameter of the Moon as seen from Earth, Deimos only one-ninth. For long periods when above the horizon they would be eclipsed by the shadow of Mars, and with Deimos the phases would be hard to make out with the naked eye. They could cause no solar eclipses, but they would often pass in transit across the face of the Sun; Phobos would do so 1300 times each Martian year, taking twenty seconds to cross the disk. Because their orbits lie in the plane of the planet's equator, they would never rise over high latitudes on the surface of Mars.

Obviously these tiny bodies are faint, quite apart from being drowned in the glare of Mars. Using an occulting bar eyepiece with my 15-inch reflector I can see them both under good conditions, but not easily. One day they may be used as natural space-stations, but of course their gravitational pulls are negligible, so that 'landing' there would be in the nature of a docking operation.

Some people have felt that Mars is a disappointment: where we had hoped to find a living world, with extensive vegetation tracts and possibly even underground water supplies, we have found a volcanic waste. Yet it is true to say that Mars is much less unfriendly than any other world in the Solar System apart from the Earth, and it must certainly be our next target for manned exploration. I am quite prepared to believe that a Martian base will be set up during the first half of the coming century, and it is quite likely that 'the first man on Mars' has already been born.

CHAPTER TEN

The Minor Planets

Any plan of the Solar System shows the huge gap between the orbits of Mars, outermost of the terrestrial planets, and Jupiter, first of the giants. Kepler was struck by it, and suspected that it might not be empty. He went so far as to write: 'Between Mars and Jupiter I put a planet.'

He knew that if such a planet existed it could not be large, as otherwise it would have been visible with the naked eye. For over a century nothing was done, but the problem was raised once more in 1772 by a German astronomer named Johann Elert Bode. Somewhat earlier another German, Titius, had discovered a curious numerical relationship linking the distances of the planets from the Sun. Bode publicized it, and nowadays it is known, rather unfairly, as Bode's Law. Whether it has any real significance is dubious. Personally I put it down to sheer coincidence, but others disagree. In any case, here it is:

Take the numbers 0, 3, 6, 12, 24, 48, 96, 192 and 384, each of which (after 3) is double its predecessor. Now add 4 to each, giving 4, 7, 10, 16, 28, 52, 100, 196 and 388. Taking the Earth's distance from the Sun as 10, the series gives the distances of the remaining planets, to scale, with reasonable accuracy (see opposite).

The three outer planets were not known when Titius worked out the relationship, but when Uranus was discovered, in 1781, it fitted well into the general scheme. Neptune, of course, is a problem. According to Bode's Law it ought not to be there, and the last figure (388) corresponds to the mean distance of Pluto, but this has no significance at all in view of Pluto's maverick nature. It is hard to avoid the conclusion that we are playing a sort of 'take-away-the-number-you-first-thought-of' game. Yet for many years the Law seemed to be very precise – except for the missing planet corresponding to Number 28.

PLANET	DISTANCE BY BODE'S LAW	ACTUAL DISTANCE
Mercury	4	3.9
Venus	7	7.2
Earth	10	10
Mars	16	15.2
—	28	—
Jupiter	52	52
Saturn	100	95.4
Uranus	196	191.8
Neptune	—	300.7
Pluto	388	394.6

In 1800 six astronomers assembled at the little German town of Lilienthal, near Bremen, where the hard-working amateur Johann Schröter had his observatory. They nicknamed themselves the 'celestial police', and determined to make a serious effort to track down the missing planet. With Schröter as President, and Baron Franz Xavier von Zach as Secretary, they worked out a scheme in which each member would be responsible for a particular region of the ecliptic – that is to say the region where the planet would probably lie, assuming that it existed at all.

A plan of this sort takes some time to bring into working order, and before Schröter's 'police' were fully organized they were forestalled. Giuseppe Piazzi, director of the Palermo Observatory in Sicily, was compiling a new star catalogue, and on 1 January 1801 – the first day of the new century* – he picked up a starlike object which behaved in a most unstarlike manner; it showed definite motion over a period of a few hours. Piazzi at first took it to be a tail-less comet, but he went so far as to write to von Zach, so that evidently he had his suspicions. Postal services were slow and unreliable, and by the time that von Zach received Piazzi's letter the object had been lost in the evening twilight.

Fortunately, Piazzi had made enough observations to enable an orbit to be worked out, and the great mathematician Gauss,

* This is really so. The last day of the old century was 31 December 1800. No doubt there will be confusion in a few years' time; the first day of the new century will be 1 January 2001, not 1 January 2000. The reason for this is that according to our time-reckoning there was no year 0. 1 BC went straight into AD 1.

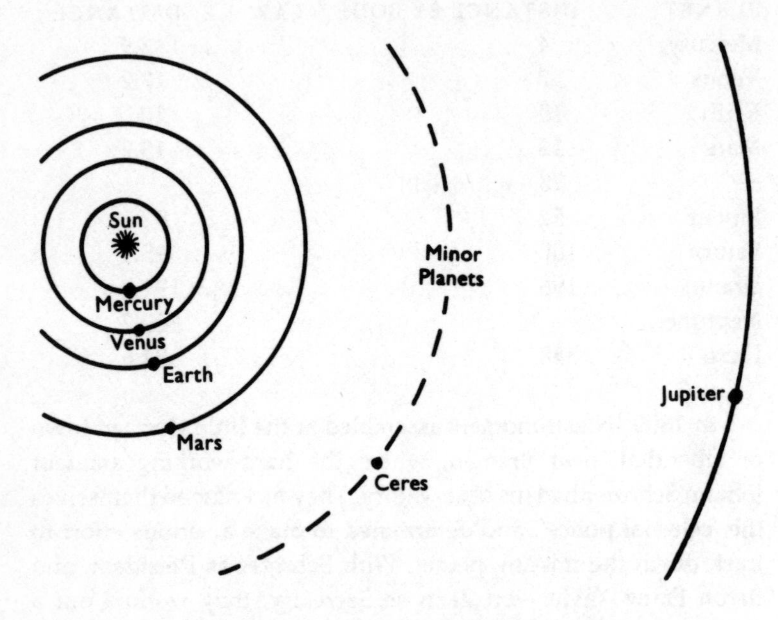

Position in the Solar System of Ceres, much the most massive of the asteroids. It fits in well with Bode's Law!

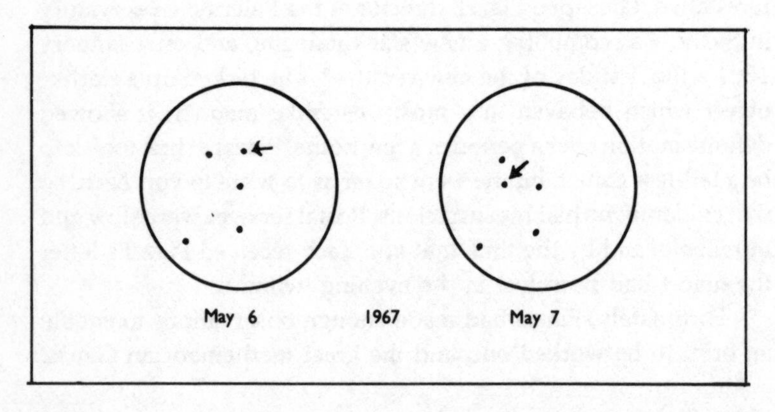

Shift in position of Ceres in 24 hours; I was using a magnification of 100 on my 3-inch refractor.

who tackled the calculations, soon saw that he was dealing with a planetary body rather than a comet. It was found again by one of the 'police', exactly a year after its original discovery, and Piazzi named it Ceres, in honour of the patron goddess of Sicily. It was found to have a distance of 27.7 on the Bode scale, which was in excellent agreement with the predicted 28; the real distance from the Sun is 257,000,000 miles, with a period of 4.6 years. The Solar System was apparently complete.

Yet the celestial police were not so sure. Ceres, less than 600 miles across, seemed hardly worthy to be ranked with the other planets, and so Schröter, von Zach and their colleagues pressed on with the search. It came as no real surprise when Olbers picked up a second small planet in March 1802. Pallas, as it was named, was so like Ceres in size and distance that Olbers believed the two to have been formed from one larger body which had met with disaster. The idea was attractive; if there were two fragments, there might be more – and so it proved. Juno was discovered by Schröter's assistant, Karl Harding, in 1804, and Vesta by Olbers in 1807.

Juno and Vesta resembled the first two members of the group, and the four became known as the Minor Planets or asteroids. No more seemed to be forthcoming, and the celestial police disbanded in 1815. Schröter himself died in the following year.

There matters rested until 1830, when a Prussian amateur named Hencke – postmaster of the little town of Driessen – took up the problem and began a systematic search for new asteroids. Alone and unaided, he worked away for fifteen years, and at last he had his reward – in the shape of a fifth minor planet, now named Astræa, circling the Sun at a distance slightly greater than Vesta's, slightly less than Juno's. However, Astræa was considerably fainter than the original four, and its diameter is less than 80 miles.

Even the enthusiastic Hencke would have been surprised to learn that his discovery was a mere prelude to thousands more. He found another asteroid, Hebe, in 1847; in the same year Hind, in London, discovered Iris and Flora; 1848 and 1849 yielded one asteroid each, and since then every year has produced its quota. By 1870 the total number of known asteroids was 109, and twenty years later it had grown to 300. Then, in 1891, Max Wolf introduced a new method which led to a rapid increase in numbers.

Wolf's method was a photographic one. If a camera is adjusted so as to follow the ordinary stars in their movement from east to west across the sky, an asteroid will show up as a streak on the plate, because it moves against the starry background quickly enough for its shift to be noticeable with a time-exposure of only an hour or two. If I set my camera to photograph a garden, and then walk in front of the lens during a time-exposure, I will appear as a blur, because of my movement. The asteroid will not blur, because it is a hard, sharp point of light, but its movement will certainly betray it.

The Wolf method was almost embarrassingly successful, and the numbers of known asteroids increased by leaps and bounds. Wolf was personally responsible for adding over two hundred, and by the beginning of 1993 there were more than 5000 asteroids with properly worked-out paths. At least a thousand more have been found on photographic plates without having been under observation for long enough to enable their orbits to be computed.

It cannot be said that at first the asteroids were popular members of the Solar System. Plates exposed for quite different reasons were often found to be swarming with short tracks, all of which had to be identified and which wasted an enormous amount of time. One irritated German even referred to the '*Kleineplanetenplage*' (minor planet pest), while an American went so far as to call them 'vermin of the skies'. Keeping track of them was very difficult – and even in the present computer age, it still is. Only one numbered asteroid has been lost (719 Albert, which is a special case), though asteroid 330 Adalberta, never existed at all; it was recorded by Max Wolf in 1892, but was photographed only twice, and it was then found that the images were of two separate stars. The last asteroid to be recovered after having 'gone missing' for many years was 878 Mildred. It was discovered in 1916, but was then lost, and turned up again only in April 1990.

Another problem was to find names for them. As the numbers grew, dignified mythological names such as Psyche, Thetis, Circe and Melpomene began to give out. The first departure was with No. 25, Phocæa, named by Benjamin Valz after a seaport in Ionia;

the next case was 45 Eugenia, honouring the wife of the French emperor Napoleon III. Some of the later names are tongue-twisting; we have 678 Fredegundis, 989 Schwassmannia, 1259 Ogylla, 1286 Banachiewicza, and so on. No. 724 is Hapag, the initials of a German navigation line, the Hamburg Amerika Paketfahre Aktien Gesellschaft, while 674 is Ekard – the word 'Drake' spelled backwards – it was named by two members of Drake University in America. No. 518, Halawe, takes its name from the favourite dessert of its discoverer, R. S. Dugan, who had developed a liking for the Arab sweet *halawe*. I particularly like No. 1372, Haremari, which was named by Karl Reinmuth in recognition of the female members (or harem!) of the astronomical institute of which he was Director. One name was actually sold. Asteroid 250 was discovered by Palisa, Director of the Vienna Observatory, who wanted to raise funds for an eclipse expedition. He therefore announced that he would sell the honour of naming his asteroid for the sum of £50. The offer was taken up by Baron Albert von Rothschild, who chose his wife's name, Bettina.

There is also No. 2309, Mr Spock. I would like to say that this was named after the sharp-eared Vulcanian astronaut of the star-ship *Enterprise*, but in fact it was named for a ginger cat which had itself been named after Mr Spock (I hope you can follow this). Finally, I cannot resist adding that No. 2602, discovered by Dr Edward Bowell in America, has been named 'Moore' after me, though I admit that it is too faint for me to see with the 15-inch telescope at my observatory.

Ceres, Pallas, Juno and Vesta are known as the Big Four, but in fact Juno is only fifteenth in size among the main-belt asteroids; the largest are Ceres (584 miles), Vesta (358), Pallas (mean 326 – it is triaxial), Hygeia (267), Davida (239) and Interamnia (210). All the rest are less than 200 miles across; Juno is spheroidal, measuring 179 × 143 miles. It is however the fifth brightest member of the swarm, because it lies in the inner part of the main zone and is fairly reflective.

Asteroids seem to fall into a few fairly well-defined types. Classes C, S and M make up most of them. Type C bodies are rich in carbon compounds; S are siliceous, and M are metallic. Juno is of type S, the most reflective class, while Ceres is of type C. The brightest of the asteroids is Vesta, which has an unusual surface

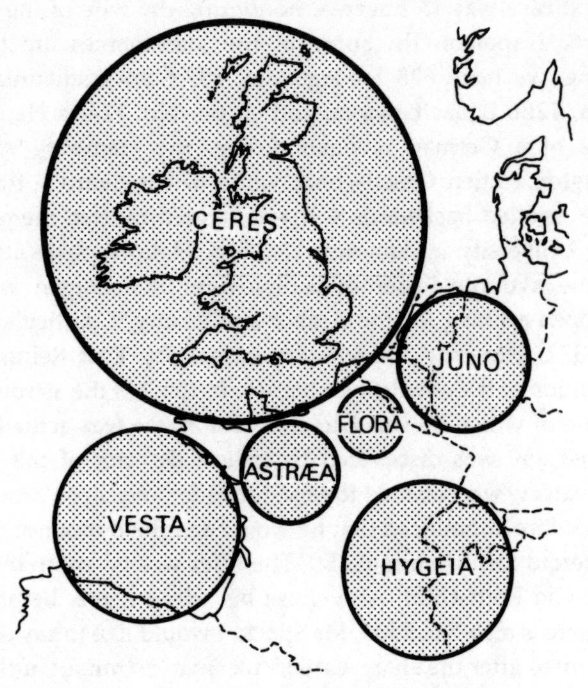

Sizes of some asteroids, compared with the British Isles.

apparently covered with igneous rock. Its mean distance from the Sun is 219,300,000 miles, closer in than Ceres, and it is the only minor planet to be visible with the naked eye. Keen-sighted people can find it easily, and I have seen it often enough, though I am sure that I would not have noticed it had I not known that it was there. Ceres, Pallas, Juno and other asteroids are within binocular range, though of course they look exactly like stars, and are identifiable only because of their movement from night to night.

No asteroid is massive enough to retain any trace of atmosphere. Ceres is very much the senior member of the swarm, and it, Pallas and Vesta make up 55 per cent of the total mass; all the asteroids combined make up only one and a half times the mass of Ceres, and the grand total is no more than 3 per cent of the mass of the Moon. Most of them are mere irregular lumps of material. The total number of asteroids is uncertain, but there is no doubt that it amounts to several tens of thousands.

In 1857 the American astronomer Daniel Kirkwood pointed out that there are definite clusters or families of asteroids, and that there are gaps in the main zone where very few bodies are found. There is no mystery about this; the tremendous pull of Jupiter is responsible. One family, of which the senior member is Flora, contains over 150 asteroids, while the more distant Hilda family accounts for at least thirty. No. 279, Thule, is unusually remote, and circles the Sun at a distance of almost 400,000,000 miles in a period of over eight years. But all the early asteroids kept strictly to the gap between the orbits of Mars and Jupiter,* and nobody was prepared for the odd orbit of No. 433, Eros, which was discovered in 1898 by Witt at Berlin.

Eros is small and never bright, but its orbit swings it well inside the main group. It has an eccentric path, and at aphelion it is well beyond Mars, but at perihelion it can come within 14 million miles of the Earth; the last two close approaches were those of 1931 and 1975. When at its nearest, its position can be measured very precisely, and its distance can be found; this gives a key to the whole scale of the Solar System. Hundreds of photographic measurements made of it in 1931 were analysed by Sir Harold Spencer Jones, who arrived at a value of 93,003,000 miles for the length of the astronomical unit or Earth–Sun distance. This is now known to be too large, and much better methods have been developed, so that in this respect Eros has lost its importance. It has a rotation period of 5.3 hours, and is irregular in shape; it is about 18 miles long by 5 miles wide – not unlike a sausage.

For some time Eros was thought to be unique, but in 1911 Palisa, at Vienna, picked up a tiny body which can approach us within 20,000,000 miles, though at aphelion it moves out into the main belt. It was numbered 719, and named Albert. Unfortunately it is only about 3 miles in diameter, and after its brief visit in 1911 it was lost; so far it has not been recovered. At the moment we simply do not know where it is.

No. 887, Alinda, discovered by Max Wolf in 1918, and No. 1036, Ganymed, found by Walter Baade in 1924, also proved to have Albert-type orbits. Then, in 1936, came the discoveries of the

* To be strictly accurate, this is not quite true of No. 132, Æthra, discovered by Watson in 1873. At its closest to the Sun its distance is very slightly less than that of Mars at aphelion – but this hardly counts! Æthra is 24 miles in diameter.

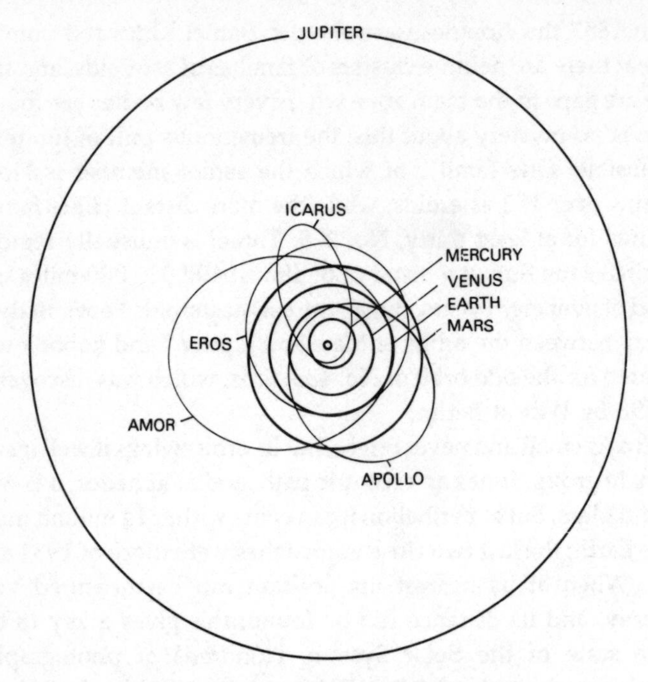

Orbits of some close approach asteroids.

first two real 'Earth-grazers' which have given their names to whole classes of asteroids; 1221 Amor, and 1862 Apollo.

Amor, no more than a mile across, was discovered by the Belgian astronomer Delporte. It came within 10,000,000 miles of the Earth, and was under observation for long enough for its orbit to be worked out. It has a period of 975 days, and after it had been twice round the Sun unseen it was picked up again in 1940. Since then it has not been lost.

Amor's reign as a record-holder was brief. Apollo, discovered by Reinmuth at Heidelberg later in 1932, can approach us to within 7,000,000 miles; at perihelion it moves inside the orbit of Venus. In theory it could transit the Sun's disk, but as it is only a mile wide there would be no chance of our being able to see it! Adonis, discovered by Delporte in 1936, veered by at a mere 1,300,000 miles. A year later Reinmuth found an even more interesting Earth-grazer, Hermes. Even smaller than Adonis, it passed by at only 485,000 miles, barely double that of the Moon.

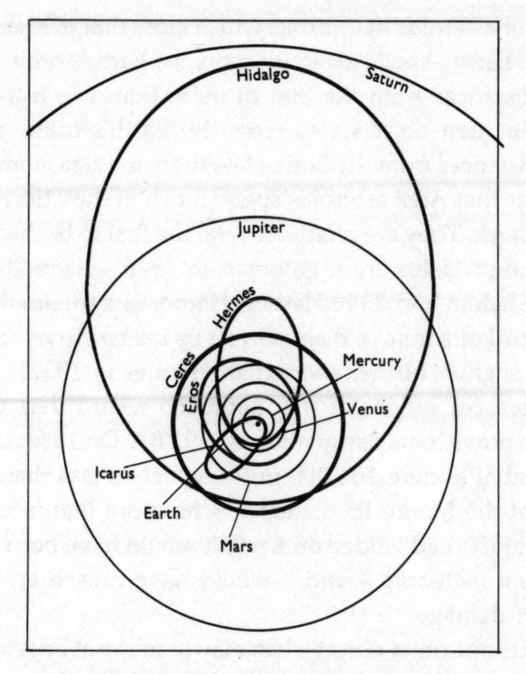

Orbits of some interesting asteroids.

When the news was made known, in January 1938, the press made the most of it, and the newspaper headlines on 10 January were sensational. 'World Disaster Missed by Five Hours', was one example. 'Scientists Watch a Planet Hurtling Earthward.' Hermes was no threat; unhappily it has been lost – and so it remains the only asteroid with a name but no official number.

Over the past few years it has become evident that close-approach asteroids are much commoner than has been believed. A systematic hunt, carried out by Eleanor Helin, Eugene Shoemaker and Caroyln Shoemaker in America, has revealed dozens of them, and more are being found every year. Few are of appreciable size; the largest, 1036 Ganymed,* has a diameter of about 25 miles, but its orbit does not actually cross that of the Earth.

Asteroids which invade the inner part of the Solar System are now divided into three classes, named after Amor, Apollo and

* Do not confuse No. 1036 with Jupiter's satellite Ganymede. It is a pity that the names are so alike.

Aten. Amor asteroids have orbits which cross that of Mars, but not that of the Earth. Apollo asteroids cross the Earth's orbit, but have average distances from the Sun of more than one astronomical unit, while Aten objects also cross the Earth's orbit, and have average distances from the Sun of less than one astronomical unit. This means that Aten asteroids spend much of their time closer in than the Earth. They are relatively rare; the first to be discovered – all by Eleanor Helin, from Palomar, in 1976 – were 2062 Aten, 2100 Ra-Shalom and 2340 Hathor. Hathor is a true midget, only about a third of a mile in diameter. At its nearest to us, in 1976, it was only slightly further away than Hermes in 1937.

The present holder of the 'approach record' has yet to be named; its provisional designation is 1991 BA. On 18 January 1991 it flew past at a mere 106,000 miles, which is less than half the distance of the Moon. Its diameter is no more than about thirty feet, so that if it had landed on Earth it would have been officially classed as a meteorite – and it would have caused an immense amount of damage.

There is one other object whch may or may not be asteroidal. It was found on 5 December 1991 by J. V. Scotti during a routine photographic search; it passed by at 286,000 miles, and from its faintness it cannot be more than around twenty feet across. There are doubts as to whether it was a genuine member of the Solar System, or simply a piece of long-lost space débris!

If these close-approach asteroids are so numerous, what are the chances of our being hit? There seems every reason to suppose that this does happen now and then, and with a major impact the whole climate might be altered, with tremendous amounts of dust flung into the upper air. As we have seen, such an event is often linked with the disappearance of the dinosaurs, 65,000,000 years ago. A surprisingly large amount of the element iridium is found in rocks laid down at that time, and iridium is a characteristic of meteorites; but everything seems to hinge upon whether the dinosaurs died out abruptly, or more gradually over a million years or so. Moreover, there have been even greater general extinctions in the past (one, for example, at the end of the Permian Period), and no similar traces of iridium have been found at these levels. I admit to being sceptical about the collision theory, but certainly it deserves to be taken very seriously.

There have even been plans to divert or destroy an asteroid which is clearly on a collision course. Breaking the object up would probably be of little use, as the various fragments would still hit us, but changing its orbit by exploding a nuclear bomb on or near it is a reasonable alternative, and could be tried. The main problem is that the tiny asteroids are often undetected until they are almost upon us, so that we might not have time to take decisive action. There is no point in being alarmist, but the possibility of a disastrous impact does have to be borne in mind.

Two known asteroids approach the Sun within the orbit of Mercury. The first, 1566 Icarus, was found by Walter Baade in 1949, when it was 8,000,000 miles away; it can never approach us closer than half this distance, so that it is not a genuine 'Earth-grazer'. At perihelion it approaches the Sun to a distance of 17,000,000 miles, much closer in than Mercury, so that it must be red-hot. At aphelion it recedes to 183,000,000 miles, well beyond the path of Mars, and will be bitterly cold. Its revolution period is 409 days. It must have just about the most uncomfortable climate in the Solar System.

Icarus paid us a visit in the summer of 1968, when it was picked up by radar. For some strange reason, reports originating from Australia caused considerable uneasiness among non-astronomers, and the popular press went so far as to warn everybody of an impending collision. The report was not officially denied for some time, simply because astronomers did not think it worth bothering about – and before a statement was made, a question about it had been asked in the House of Commons!

Icarus is named after the mythological youth who took to the air by using artificial wings, and met an untimely death when he flew too near the Sun, so that the wax fastening his wings melted. The strange little world can by-pass all the inner planets; for instance, on 1 May 1968 it was within 10,000,000 miles of Mercury.

The second 'Sun-grazer', 3200 Phæthon, was discovered in 1983. It can swing even closer in than Icarus, with a minimum distance from the Sun of 13,000,000 miles, though at aphelion it moves out to 118,000,000 miles, well beyond the orbit of the Earth; the revolution period is 522 days, and like Icarus it is tiny, with an estimated diameter of three miles. But in one way

Phæthon is immensely significant. It moves in a path very similar to that of the Geminid meteor stream, which produces a shower of shooting-stars seen regularly every December.

It has long been known that meteors are cometary débris. Therefore, is it possible that Phæthon used to be a comet, and laid the dusty Geminid trail before losing all its volatiles? There had been many suggestions of a link between comets and close-approach asteroids, and it had been claimed that the two classes of objects were one and the same. Further evidence came to light when it was proved that Asteroid 4015, discovered by Mrs Helin in 1979, is identical with Comet Wilson-Harrington, which had been lost ever since 1949. It used to have a tail; it has none now, and it is starting to look as though some of the 'Earth-grazers', at least, are nothing more nor less than dead comets, now reduced to their inert nuclei.

For many years it was thought that the most remote member of the asteroid swarm was 279 Thule, discovered as long ago as 1888. It has a fairly circular orbit, taking it from 392,000,000 miles to 397,000,000 miles in a period of 4.2 years, and by asteroidal standards it is rather large, with a diameter of around 80 miles. Then, in 1906, Max Wolf at Heidelberg detected 588 Achilles, which was soon found to be much more remote. It moves in the same orbit as Jupiter, so providing mathematicians with an interesting demonstration of a Lagrangian point.

I have referred to Lagrangian points before, when discussing the reported 'Kordylewski clouds' in the same orbit as the Moon. It was as early as 1772 that the famous French mathematician Joseph Lagrange called attention to the special 'problem of three bodies', which arises when a massive planet and a tiny asteroid move round the Sun in the same plane, in circular orbits and with equal periods. Lagrange found that if the bodies are 60 degrees apart, they will always remain 60 degrees apart. Achilles behaves in just this way. Subsequently other similar asteroids were found, and were given the names of combatants in the war between Greece and Troy, so that nowadays they are known collectively as the Trojans. Dozens of them are known, and some are well over a hundred miles across, but their remoteness makes them look faint.

The Trojans do not keep strictly 60 degrees ahead of or 60 degrees behind Jupiter, because we are dealing with elliptical orbits, and there are various other factors to be taken into account – notably perturbations caused by Saturn. For example, 1437 Diomedes may wander as far as 40 degrees beyond the Lagrangian point on the side away from Jupiter, and 24 degrees from it on the side toward Jupiter. (There is, incidentally, one known 'Martian Trojan', 1990 MB, discovered in that year by astronomers at Kitt Peak in Arizona. No doubt others exist.)

Asteroid 944 Hidalgo, discovered by Walter Baade in 1920, is not a Trojan, but has an orbit which carries it from the inner part of the Solar System out to well beyond Jupiter in a period of fourteen years. Even more extreme is 5145 Pholus, discovered in 1992; the distance from the Sun ranges between 840,000,000 miles and

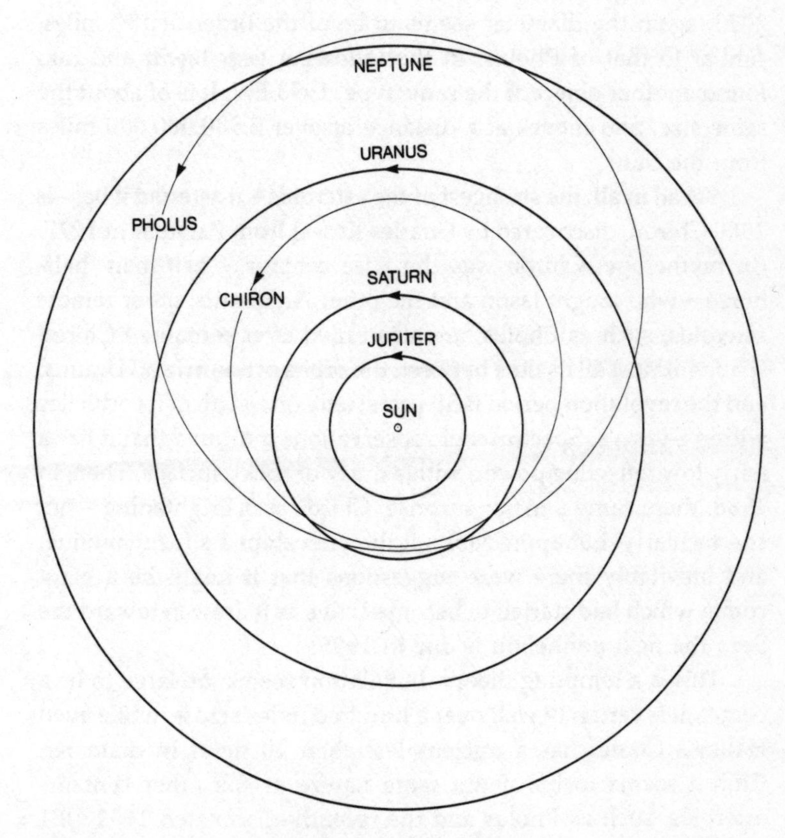

Orbits of Chiron and Pholus.

2,976,000,000 miles, so that Pholus' orbit crosses those of Saturn, Uranus and Neptune. The revolution period is 93 years, and the tilt of the orbit is as much as 25 degrees, so that Pholus moves well away from the elliptic. It appears to be reddish – there are suggestions that it may be coated with organic material – and the diameter may be around 150 miles, which by asteroidal standards is large.

However, even Pholus has been outmatched by 1992 QB1, found in that year by David Jewitt and Jane Luu. At the time of its discovery it was moving well beyond Neptune, and has been found to have a path which takes it out as far as 4,126,000,000 miles from the Sun, though at perihelion it comes in to 3,685,000,000 miles – not far from the orbit of Neptune. The orbital period is 295.7 years, and the next perihelion is due in 2023; again the diameter seems to be of the order of 150 miles, similar to that of Pholus. In the following year Jewitt and Luu found another object of the same type, 1993 FW. It is of about the same size, and moves at a distance of over 3,900,000,000 miles from the Sun.

Yet all in all, the strangest of the asteroids – if asteroid it be! – is 2060 Chiron, discovered by Charles Kowal from Palomar in 1977. (In mythology, Chiron was the wise centaur – half-man, half-horse – who taught Jason and the other Argonauts; other remote asteroids, such as Pholus, are also called after centaurs.) Chiron spends almost all its time between the orbits of Saturn and Uranus, and the revolution period is 50 years; only one-sixth of its orbit lies within Saturn's. Spectroscopic observations indicated that it has a fairly low reflecting power, with a dusty or rocky surface. Then, in 1988, there came a major surprise: Chiron was brightening – not spectacularly, but appreciably. It then developed a fuzzy outline, and inevitably there were suggestions that it might be a giant comet which had started to become active as it drew in toward the Sun; the next perihelion is due in 1995.

This is a tempting theory, but Chiron seems too large to be a comet. It is certainly well over a hundred miles across, while even Halley's Comet has a nucleus less than 20 miles in diameter. Chiron seems to be of the same nature as the other centaur-asteroids, such as Pholus and the recently-discovered 1992 QB1 and 1993 FW.

There seems a good chance that the 'centaurs' are planetesimals, material left over when the main planets were formed from the solar nebula. Most of them have unstable orbits. It is worth noting that in 1664 BC Chiron approached Saturn to within 10,000,000 miles, and this is not much greater than the distance between Saturn and its outermost satellite, Phœbe, which has retrograde motion and is almost certainly an ex-asteroid about the same size as Chiron.

It is generally thought that comets come from a whole reservoir of icy bodies orbiting the Sun at a distance of about one light-year or six million million miles; this is termed the Oort Cloud in honour of the late Jan Oort, who first proposed its existence. More recently it has been suggested that there may be an inner reservoir, the Kuiper Cloud, not far beyond the orbit of Neptune. If so, then the centaur asteroids may be the first-detected members of the Kuiper Cloud.

The distribution of the main-belt asteroids shows that there are well-defined 'families' – that of Flora has over a hundred and fifty members -- and there are also gaps where few bodies are found; these were predicted in 1857 by Daniel Kirkwood, and are named Kirkwood Gaps after him. There is no mystery about them; they are caused by the powerful perturbing influence of Jupiter. When an asteroid moves into one of the gaps, Jupiter gradually moves it out again. It also seems that Jupiter is directly responsible for the existence of the swarm. No large planet could form in this region of the Solar System; every time a major body started to build up, Jupiter literally tore it apart, so that the end product was a swarm of dwarfs. An older idea, that the asteroids are the remnants of a former planet or planets which met with disaster and broke up, has been more or less abandoned.

At least we now have reliable information about one main-belt asteroid, 951 Gaspra, which has been photographed from close range by the spacecraft Galileo, *en route* for Jupiter. The pictures, taken on 13 November 1991, show that Gaspra is wedge-shaped (not unlike a distorted potato) with a darkish, rocky, crater-pitted surface; it measures 12 × 7 miles, so that in size it is not very different from the nucleus of Halley's Comet. It gives every indication of being a fragment of a larger body which has broken up, and indeed collisions in the asteroid belt must have been common in the past; possibly they still are. The main surprise about Gaspra is that it seems to have an appreciable magnetic field.

Telescopically, asteroids look just like stars, and they can be identified only by their movements from night to night. Yet there is useful work to be done by the well-equipped amateur. For example, careful measurements of their changes in brightness give clues about their rotation periods, which show a wide range, from 2 hours 16 minutes for Icarus up to about 1500 hours for 288 Glauke. For this, a photometer is more or less essential, though really careful eye estimates cannot be discounted if there are any suitable comparison stars nearby. Even more important are occultations of stars by asteroids, because the duration of the occultation depends on the size and shape of the asteroid. Amateurs can be of particular help here, because the track of an occultation is narrow and always uncertain – and only the amateur is likely to be able to take portable equipment to the critical position.

Junior members of the Sun's family though they may be, the asteroids are far from unimportant, and at the present time they are being closely studied by both professional and amateur astronomers. Few people now agree with the contemptuous dismissal of them as 'vermin of the skies'.

CHAPTER ELEVEN

Jupiter

Far beyond the main asteroid zone we come to Jupiter, giant of the Solar System. It moves round the Sun at a mean distance of 483,000,000 miles, and has a 'year' almost twelve times as long as ours. There can be no proper seasons, because the axis of rotation is almost perpendicular to the orbital plane, but in any case seasons would be meaningless on a world such as Jupiter.

The orbital velocity is less than half that of the Earth; the synodic period is 399 days, so that Jupiter comes to opposition every year, and there are several months annually when it is well placed for observation. It is outshone only by Venus and, very occasionally, by Mars.

Jupiter and the Earth compared.

Jupiter owes its brilliance mainly to its immense size. Measured through the equator, the diameter is 89,420 miles, but the polar diameter is less, because – as any telescope will show – the globe is appreciably flattened. This is because of the rapid rotation. Jupiter's 'day' is shorter than for any other planet, and at the equator the period is only 9 hours 50½ minutes, so that particles there are being whirled along at 28,000 mph. Away from the equatorial zone the rotation period is five minutes longer, and different surface features have periods of their own, so that they drift around in longitude.

Jupiter is not 'solid' in the usual meaning of the term; the surface is gaseous, and all we are seeing are the upper clouds. Overall the density is low, and though Jupiter has a volume 1300 times greater than that of the Earth it 'weighs' only 318 times as much as the Earth. Even so, Jupiter's mass is greater than those of all the other planets put together, and it has been said that the Solar System consists of the Sun, Jupiter and assorted débris!

The clouds we can see are very cold indeed, but at its core Jupiter must be hot; estimates range between 30,000 and 50,000 degrees Centigrade (54,000 to 90,000 Fahrenheit). Moreover, Jupiter sends out 1.7 times as much energy as it would do if it depended entirely upon what it receives from the Sun. Can it be a 'minor star' in its own right?

This was once a very popular idea, and little more than a century ago R. A. Proctor could write: 'Jupiter is still a glowing mass, fluid probably throughout, still bubbling and seething with the intensity of the primæval fires, sending up continuous enormous masses of clouds, to be gathered into bands under the influence of the swift rotation of the giant planet.' It is a fascinating picture, but we now know that there is an essential difference between a planet and a star. The distinction hinges chiefly on mass – and remember, it would take over a thousand Jupiters to make up one body as massive as the Sun.

To recapitulate: a star is formed out of dust and gas. The material accumulates because of gravitational forces; as the particles move inward towards the centre of the contracting cloud, collisions make the density and the temperature rise. When the temperature has reached a critical value of around 10,000,000 degrees Centigrade (18,000,000 degrees Fahrenheit), nuclear reactions are triggered off, and the star starts to shine. For this to happen, it seems that the embryo star must be at least ten times as massive as Jupiter is today. Therefore, Jupiter's core can never have become hot enough for nuclear reactions to begin. Recently we have tracked down bodies, far away in space, which are almost on the limit between planets and stars; they are known, rather misleadingly, as Brown Dwarfs. I will have more to say about them later. But Jupiter is not massive enough even to be classed as a Brown Dwarf.

There are two theories to account for the high core temperature. It may be that Jupiter is slowly shrinking, at the rate of around one millimetre per year, so that the excess energy is gravitational; but on the whole it seems more likely that what we find is simply the heat left over, so to speak, when Jupiter was formed from the solar nebula. It has not had enough time to cool down.

Look at Jupiter through a telescope, and you will see that the yellowish, flattened disk is crossed by dark belts; there are bright intermediate zones, and various other features such as spots, wisps and festoons which are always changing. Watch for a few minutes, and you will see the features being carried slowly across the disk from one side to the other. It takes less than six hours for a feature to be carried from one limb across to the other.

The first step in trying to understand the nature of Jupiter was taken in the 1930s, by spectroscopic observations. Hydrogen and helium would be expected to be plentiful, just as they are in the Sun, but both these gases are rather shy of showing themselves under conditions of the sort we find on Jupiter, so that the first positive identifications were of compounds: ammonia and methane. Ammonia has the chemical formula NH_4, so that its molecule is made up of four hydrogen atoms together with one atom of nitrogen; methane, better known as marsh gas, has the formula CH_4, so that it is a combination of hydrogen and carbon. (Miners dread methane, which is explosive; they call it firedamp.) Since then we have been able to make more accurate measurements. Jupiter's atmosphere is 89 per cent hydrogen; helium accounts for another 11 per cent, and this leaves a mere one per cent for all other elements. But what about the interior?

In 1932 Rupert Wildt proposed that there was a rocky core,

overlaid by a thick layer of ice which was in turn overlaid by the hydrogen-rich atmosphere. Next came a theory due to W. R. Ramsey, who believed that the core too was made up of hydrogen, so compressed that it took on the characteristics of a metal; above the core was a deep layer of ordinary solid hydrogen, covered by the atmosphere.

In some ways Ramsey's theory was not too wide of the mark, but we now have a fairly reliable picture, showing that the hot, iron silicate core is surrounded by shells of liquid hydrogen – the lower shell being metallic, and the outer shell ordinary molecular hydrogen. The gaseous atmosphere above the top of the liquid is 600 miles deep. In it there are various definite cloud layers, including one layer of water-ice crystals; there may even be a layer of liquid water. The uppermost regions are characterized by crystals of ammonia, known generally as ammonia cirrus. The vivid hues seen on Jupiter must be due to chemical and physical processes (lightning, for example) which add colour to the basically white ammonia cirrus.

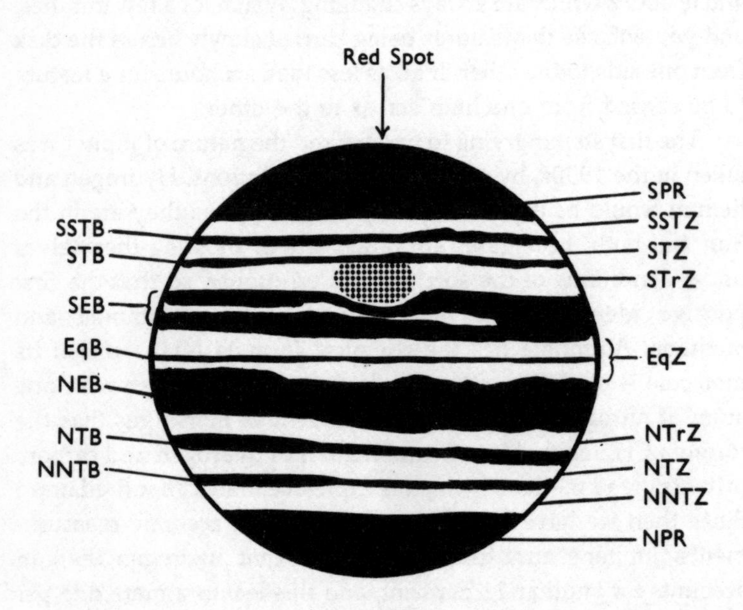

Nomenclature of Jupiter. N = north; S = south. T = temperate, Tr = tropical, Eq = equatorial, P = polar, R = region, Z = zone, B = belt.

The most prominent markings on the disk are the belts, which run parallel to the planet's equator. They are always very much in evidence, and generally speaking their latitudes do not change much. The accepted Jupiter nomenclature is given in the diagram; I have adopted the usual astronomical practice of putting south at the top, though nowadays this is not always followed.

Though the main belts are nearly always on view, they are not constant in either breadth or intensity. Generally the North Equatorial Belt (NEB) is the most conspicuous belt, but there are times – as in 1985–6 – when the South Equatorial Belt (SEB) was dominant. On the other hand the SEB is very variable. At some time during May or June 1989, when Jupiter was on the far side of the Sun, the SEB disappeared completely, but soon re-formed; at the end of 1992 both the SEB and the South Temperate Belt (STB) almost vanished. The most remarkable Jupiter year in my own experience was, I think, 1962, when the two equatorial belts merged into a solid wedge of chrome yellow extending right round the globe.

Many past observers have recorded bright colours: greens, blues and even violet. I can only say that I have never seen them myself, and they seem to have been recorded less often during the past few decades. Whether the colours on the disk really are less pronounced today, or whether older observers tended to exaggerate faint hues, is not clear.

Though the latitudes of the features do not change much, the longitudes do, and different regions have different rotation periods. System I – that is to say, the area bounded by the north edge of the South Equatorial Belt and the south edge of the North Equatorial Belt – has an average period of five minutes shorter than that of the rest of the planet (System II), so that there is a fast-moving current along the equator. Yet these figures are only a mean; special features, such as the Red Spot, drift around in longitude.

It is not hard to measure the rotation periods of special features on Jupiter. Because the spin is so quick, a few minutes' observation will show that the features are being carried across the disk, from left to right, with an ordinary inverted telescopic image. What has to be done is to time the moment when the feature reaches the central meridian. The diagram overleaf shows what is meant; the

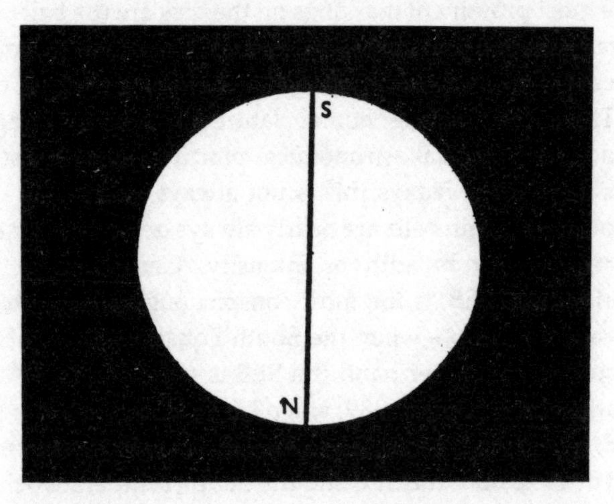

Central meridian of Jupiter – easily found, because of the obvious flattening of the globe.

central meridian is easily located, because of the flattening of the disk, and transits of the surface features can be timed to an accuracy of less than a minute. Obviously, successive transits of the same feature will give the rotation period. In practice it is seldom possible to obtain two transits of the same feature in one night, because the interval is almost ten hours; I have done it on suitable occasions when Jupiter has been high in the northern hemisphere, but it is not easy. However, the transits need not be of successive rotations.

The table on pages 134 and 135 is based on my own observations between 1951 and the present time. I do not pretend to be eagle-eyed, but at least the observations have been made by the same observer with the same equipment, so that they may serve to give a general indication of what has gone on.

Most of the interesting activity takes place near the equator and in the south tropical and south temperate zones. The poles are less turbulent, and in general the northern hemisphere is the calmer of the two. This may well be due to the presence in the south of one truly remarkable feature: the Great Red Spot.

The Spot became very conspicuous in 1878, but it was by no

means new. Robert Hooke apparently saw it in 1664; Cassini, first Director of the Paris Observatory, drew it in 1665, and it may have been seen even earlier. It is extremely long-lived, and there is nothing else like it.

In 1878 the Spot became brick-red in colour, and measured 30,000 miles long by 7000 miles wide, so that its surface area was equal to that of the Earth. After 1882 it faded, and since then it has disappeared completely at times, as from the mid-1970s to the late 1980s, but it always comes back, and its site is marked by indications of what is termed the Red Spot Hollow, a large 'bay' in the South Equatorial Belt. The colour, too, is very variable, and between 1990 and 1993 it was grey; I could not see even a tinge of redness in it.

Associated with the Red Spot in the days when I first began to study Jupiter was a most interesting feature known as the South Tropical Disturbance (STD). It took the form of a shaded zone between two white spots; it was of the same latitude as the Red Spot, and had a slightly shorter rotation period, so that every few years it caught up with the Spot and passed it. While this was happening there were marked interactions between the Disturbance and the Spot; the Disturbance was accelerated, and as it passed by it seemed to drag the Spot along with it for several thousands of miles. When the Disturbance had gone on its way, the Spot drifted back to its original position. The Disturbance was first reported in 1901; it was last observed in 1940, and seems to have vanished for good. Now and then there are indications of a revival (I saw something of the kind in 1967), but on the whole it seems that these features are not true returns of the Disturbance – and they do not last for long.

Before the Space Age there were various theories to explain the nature of the Red Spot. Some were decidedly wild (in 1943 E. Schönberg proposed that it was nothing more nor less than the red-hot top of a volcano!) but at least it was clear that the Spot was not fixed. The favoured theory was due to Bertrand Peek, an English amateur who became world-famous for his observations of Jupiter. Peek regarded the Spot as a solid body floating in the outer atmosphere of the planet, and believed that its changes in visibility were due to changes in its level; if the Spot sank, it would

SURFACE CHANGES ON JUPITER, 1951–92 8.5in, 12.5in, 15in reflectors: ×230–460.

D = very dark. d = fairly dark. B = very broad. b = fairly broad. P = very prominent. p = fairly prominent.
n = narrow. o = obscure. a = absent. R = very red. r = red. x = double. RS = Red Spot.

YEAR	NEB	SEB	NTB	STB	RS	LONGITUDE RS	COMMENTS
1951	PBd	no	n	PnD	a	–	NEB dominant.
1952	pbd	pn	n	Dn	a	–	NEB dominant.
1953	PBd	bd	n	no	a	–	SEB reviving.
1954	PBd	Bpx	o	no	a	–	EqB sometimes easy.
1955	PBd	Pbd	o	o	a	–	SEB double at times.
1956	PBD	Bdx	o	dn	a	–	At times SEB as prominent as NEB.
1957	PBD	no	bd	Pbd	P	–	SSTB sometimes prominent.
1958	PBD	pdn	no	Pbd	P	–	SEB area active.
1959	PBDx	Bdp	pd	bp	o	–	EqZ yellow at times.
1960	PBD	Pd	pd	Pdx	o	–	RS grey. STZ very bright.
1961	PBD	Bpd	pd	dn	pr	327	EqZ dusty.
1962	EqBs merged	over EqZ	o	Pd	Pr	007	EqZ chrome yellow; EqBs merged.
1963	PBd	Pdb	bo	Pdb	Pr	015	EqZ dusky; EqBs occasionally merged. EqZ active.
1964	PBd	PBd	o	Pdb	PR	019	EqBs so broad that they almost merge.
1965	PBd	PBdx	bo	pdb	PR	024	EqBs comparable.
1966	PbD	Pbx	pbx	pbx	or	026	SEB very variable.
1967	PBD	Pbx	Pb	pb	Pr	028	NEB very active.
1968	pbx	PBdx	bo	pdn	Pr	028	SEB often more prominent than NEB.
1969	PBd	Bpx	o	Bpd	PR	032	EqZ chrome yellow. EqB sometimes seen.
1970	PBd	no	bo	BPD	PR	026	SEB faint and occasionally double.

1971	PBD	Bpx	Pdx	PDn	PR	010	EqBs comparable at times.
1972	PBd	PBd	do	bdp	PR	003	EqZ very yellow at times.
1973	PBD	PBdx	pd	Dnp	PR	003	SEB occasionally superior to NEB.
1974	PBD	PBd	Pbd	pbd	PR	025	EqBs comparable.
1975	bpd	bpd	Pdb	dpn	PR	048	EqZ dusky. SEB sometimes very light.
1976	PBD	pbdx	Pbd	pdn	a	–	SEB broad, often double, sometimes faint.
1977	pbd	PBD	pbd	npd	a	–	SEB dominant.
1978	Pbd	pbdx	pn	pdn	a	–	SEB as broad as NEB, but less dark.
1979	PBd	Pbdx	no	pn	a	–	RSH = 062. SEB often brownish.
1980	PBD	pbdx	bdn	dpn	a	–	EqBs sometimes comparable.
1981	Bpb	dpb	Bdp	ndp	o	–	RS suspected as grey inside RSH.
1982	pbd	PBD	b	no	a	–	SEB more prominent than NEB.
1983	pbd	PBD	pn	no	a	–	SEB dominant.
1984	pdn	PBd	bo	no	a	–	RSH at 028.
1985	pDn	PBdx	bo	no	a	–	SEB dominant.
1986	pDn	PBD	no	nd	a	–	SEB generally superior to NEB.
1987	pDn	BPd	bdp	no	a	–	SEB much broader than NEB, but less dark.
1988	pdbx	PDbx	pdb	no	a	–	EqBs comparable.
1989	pbdx	PBP = a	bpx	no	p	026	SEB vanished between Apr. 12 and July 31!
1990	PbD	a = PBx	pdn	bd	p	032	RS grey EZ chrome yellow. SEB reappeared: dominant.
1991	PbD	bx	Pdx	no	p	033	RS grey. EqZ sometimes yellow. SEB sometimes faint.
1992	PBd	bo	Pbd	no	p	036	RS grey. SEB broad but light; then obscure.
1993	PDn	o	PDb	o	p	040	RS grey. SEB, STB barely visible.

be covered up, while if it rose to the top of the cloud-layer it would be conspicuous. Calculations showed that the total change in level would not have to be more than about seven miles.

To explain the cause of this rising and falling, Peek drew a comparison with the well-known experiment of immersing an egg in a solution of salt and water. If the solution is densest at the bottom of the jar, as will probably be the case, the egg will float at a level determined by the density of the liquid. Add some more salt, thereby increasing the density of the liquid, and the egg will rise to the top of the jar. Therefore, a slight increase in the density of Jupiter's atmosphere in the region of the Spot would force it upward.

An alternative theory was due to Raymond Hide, who maintained that if Jupiter had a fairly well-defined surface below the clouds, we could expect strong winds there (or, more accurately, strong currents). If there were some large obstruction on the surface – a mountain, if you like – the atmosphere might circulate around it, and above the obstruction there would be a column of stagnant gas, known as a Taylor Column. In this case, the Red Spot would simply be the top of the column.

Thanks to the space missions, we now know that both these theories are wrong. The Red Spot is a gigantic whirling storm – a phenomenon of Jovian 'weather'. It has lasted for so long because it is so large, but it may not persist permanently. The colour is still something of a mystery, but it may well be due to phosphorus.

There was an interesting development in 1955, when B. F. Burke and K. L. Franklin, in America, found that Jupiter is a source of radio waves concentrated in wavelengths of tens of metres (decametric emissions) and tenths of metres (decimetric). The discovery was unexpected, and was also accidental – Burke and Franklin were not looking for anything of the sort, but picked up the radio waves every time that Jupiter came into a suitable position with respect to their equipment. Let it be stressed that there was never any suggestion that the radio waves could be artificial; after all, both light-waves and radio waves are electromagnetic phenomena, and the only essential difference is in wavelength. But at first the emissions from Jupiter were very much of a puzzle. They did not show any sign of being linked with discrete features on the disk (though it took a long series of visual and radio observations to prove this – a programme in which I was involved). However, it was found that there was a connection between the radio bursts and the orbital position of Jupiter's innermost large satellite, Io. This was another surprise – at that time we had no idea that Io was actively volcanic. It began to look as though Jupiter must have a very powerful magnetic field, plus zones of radiation similar to but much stronger than the Van Allen zones encircling the Earth.

Earth-based observations of Jupiter could tell us a great deal, but what was really needed was close-range information from space vehicles. The first of these were the Pioneers – No. 10, which passed by its target in December 1973, and No. 11, which followed a year later. Both were successful, but were completely outshone by the first two Voyagers, both of which were launched in 1977 and both of which by-passed Jupiter in 1979, Voyager 1 on 5 March and its twin on 9 July. Two more probes have been sent up since then. Ulysses is designed to study the poles of the Sun, but had first to pass by Jupiter, which it did in February 1992. The spacecraft appropriately named after Galileo was launched in 1989, but will not reach the neighbourhood of Jupiter until 1995.

Pioneer 10 blazed the trail. It began its journey from Cape Canaveral on 2 March 1972; a year and nine months later it was within 82,000 miles of Jupiter, having survived the crossing of the asteroid belt (a part of the journey about which the NASA planners had considerable qualms, because nobody was sure how many small asteroids there might be, and a collision between the spacecraft and a lump of material the size of, say, a football would be disastrous). Pioneer found that there was an unforeseen hazard. Jupiter's radiation zones were much stronger than had been expected, and had the probe ventured any closer to the clouds its instruments would have been put permanently out of action. However, all was well – just! – and spectacular pictures were sent back, showing the Red Spot, the coloured regions, the belts and the zones. It was confirmed that Jupiter is a very violent world, with tremendous storms raging all the time, and that the magnetic field is very powerful indeed.

Pioneer 11 was a near-twin of its predecessor, and reached the neighbourhood of Jupiter a year later; this time the orbit was modified, because Pioneer 11 approached within 29,000 miles of the cloud-tops and had to be swept quickly over the equatorial zone, where the radiation danger was at its worst. It confirmed the earlier results, but its mission was not over; there was enough power left to swing it back across the Solar System to a rendezvous with Saturn in 1979. Neither Pioneer will ever return. They are travelling too fast, and in a few years' time we are bound to lose all track of them.

I will say no more about the Pioneers, because they were preliminary missions only, and we learned much more from the Voyagers of 1979, at distances of 217,530 miles and 443,750 miles respectively (it is truly remarkable to find that after such a long journey, distances of this kind can be given so accurately). One of the first major discoveries was that of a ring system. The Jovian rings are quite unlike those of Saturn, and are so thin and dark that there is no chance of seeing them with Earth-based telescopes. The nature of the Red Spot was finally found, and superb pictures were obtained of the entire surface – which had changed markedly since the Pioneers had passed by. Auroræ were detected, and it was found that there is almost constant thunder and lightning, so that Jupiter is a noisy world as well as a turbulent one (and, in view of the abundance of methane and ammonia, smelly as well!).

Perhaps the most important investigations of all carried out by the Voyagers concerned Jupiter's magnetic field, which is not only powerful but also very extensive; the magnetosphere is so large that at times it may even engulf Saturn. Moreover, the Voyagers confirmed the link between the magnetic field and the volcanic satellite Io.

So far I have said little about Jupiter's attendants, but they are of special interest. The main four are so large and bright that any small telescope will show them; so will good binoculars, and a few very keen-sighted people can see them with the naked eye. They were first studied in detail by Galileo, using his primitive telescope in January 1610, and are known collectively as the Galileans. Actually, they may have been seen slightly earlier by another observer, Marius, and it was Marius who suggested the names we now use: Io, Europa, Ganymede and Callisto. Io is slightly larger than our Moon, Europa slightly smaller, and Ganymede and Callisto much larger; in fact Ganymede has a diameter greater than that of Mercury, though it is not so massive.

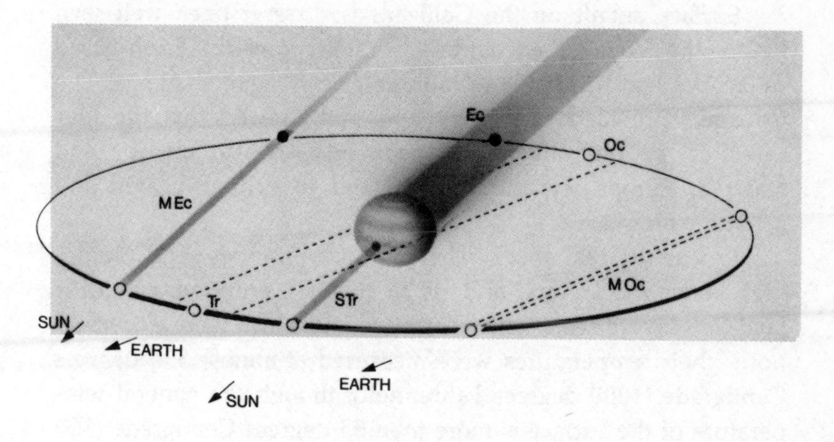

Phenomena of the Galilean satellites.

The movements of the Galileans are easy to follow from night to night; their orbital periods range between 1 day 19 hours for Io up to 16 days 18 hours for Callisto. They may pass into Jupiter's shadow, and be eclipsed; they may be occulted by Jupiter's disk; they and their shadows may pass in transit across the planet, so that they can be seen as they track slowly along – the shifts in position are evident after only a few minutes. During transits, Io and Europa are usually hard to find except when they are close to the limb, but the less reflective Ganymede and Callisto tend to show up as greyish spots. The shadows are always black. Mutual phenomena also occur; for example Io and Europa may be eclipsed by the shadow of Ganymede, and now and then Europa's shadow falls across Io; it is also possible for one satellite to occult another. These mutual phenomena are not important, but they are fascinating to watch with an adequate telescope.

The eclipses of the Galileans by the shadow of Jupiter led to a major discovery in 1675, by the Danish astronomer Ole Rømer. Rømer found that the calculated eclipse times did not agree with observation, and he realized that this is because light travels at a finite velocity; remember, Jupiter is not always at the same distance from us. After careful calculations Rømer arrived at a value for the velocity of light which was close to the true figure of 186,000 miles per second.

Surface details on the Galileans had never been well seen before the Voyager encounters, though powerful Earth-based telescopes had been able to indicate a few vague shadings. The Voyagers provided excellent maps, and showed that the four satellites are not alike. Ganymede and Callisto are icy and cratered, Europa icy and smooth, and Io sulphur-coated and actively volcanic.

Io was the real shock. The sulphur volcanoes send plumes high above the surface, and are in constant eruption; extensive lava-flows were seen, with volcanic vents which were extremely hot – their temperatures were measured at almost 550 degrees Centigrade (1000 degrees Fahrenheit), though the general temperature of the surface is more than 55 degrees Centigrade (100 degrees Fahrenheit) below zero. The materials are sent out at high speed, and Io's volcanoes are certainly more violent than ours. There are also lakes of black, liquid sulphur, in which masses of solid sulphur float rather in the manner of icebergs.

It may well be that Io's crust is a 'sea' of sulphur and sulphur dioxide between two and three miles deep, with only the uppermost half-mile solid. Heat escapes from the interior in the form of lava erupting below a sulphur ocean, and the result is the outpouring of a mixture of sulphur, sulphur dioxide gas, and sulphur dioxide 'snow'; in a way it might be better to use the term 'geyser' instead of volcano. The interior heat is thought to be due to the fact that Io's path round Jupiter is not quite circular, with the result that the globe is subject to variable 'tugs' by the gravity both of Jupiter and of the other Galileans, churning and heating it. Material sent out from the Ionian volcanoes is captured by Jupiter, and produces a doughnut-shaped torus round the planet centred on Io's orbit. Moreover, Io and Jupiter are connected by strong electric currents. We can now understand why Io has so marked an effect on Jupiter's radio waves.

One thing is certain: we will never go to Io. The unstable surface, the constant eruptions and the intense cold away from the volcanic vents are daunting enough, but in addition Io moves in the thick of Jupiter's radiation zones. It must be the most lethal world in the Solar System. At least we can now go on monitoring

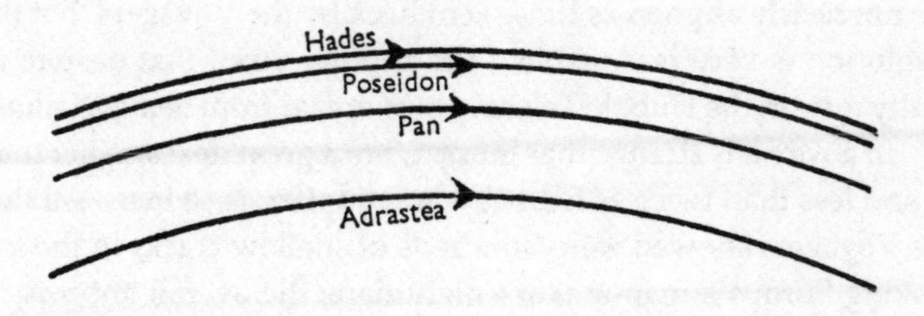

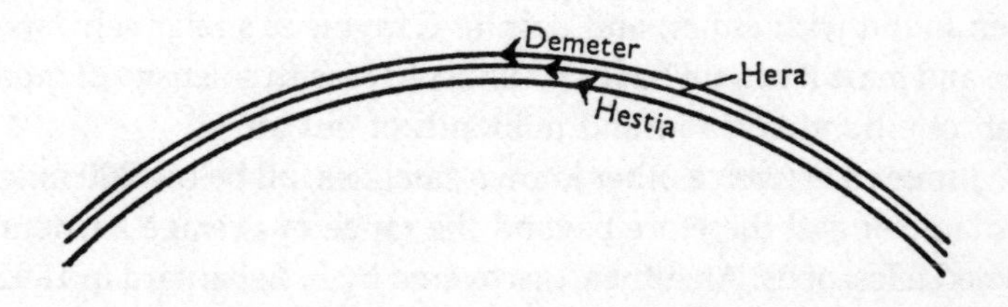

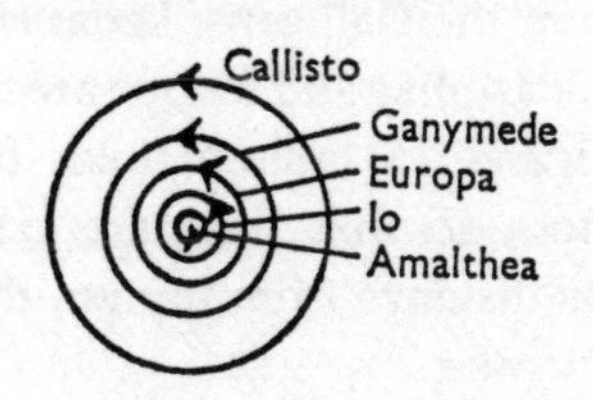

Plan of Jupiter's satellite family. This shows the scale (obviously it is impossible to put in the very small satellites closer in to Jupiter than Io) and also the retrograde movements of the outer four, but it must always be remembered that the satellites beyond Callisto are so far from Jupiter that their orbits are not even approximately circular, and do not repeat themselves exactly in successive revolutions.

some of the volcanoes, because they are shown in the pictures produced by the Hubble Space Telescope. Of course the pictures are not nearly as good as those sent back by the Voyagers, but the resolution is very reasonable – yet another proof that despite its faulty mirror, the Hubble Telescope is very far from being a failure.

In a way it is strange that Europa, not a great deal smaller than Io and less than twice as far away from Jupiter, is so inert. All that the Voyagers showed were hundreds of shallow cracks in the ice, making Europa a map-maker's nightmare; the overall impression is not unlike a cracked eggshell. It has been seriously proposed that below the icy crust there is an ocean of liquid water which could support primitive life-forms. I can only say that I take this suggestion with a very large pinch of cosmic salt.

Ganymede too has an icy crust, but this time with craters, plains and streaks; Callisto has so many craters that there seems to be almost no level ground anywhere. No trace of atmosphere has been found with either, and despite Ganymede's relatively large size and mass it cannot have an atmosphere with a density of more than one hundred thousand millionth of our air.

Jupiter has twelve other known satellites, all below 200 miles in diameter and therefore beyond the range of average amateur-owned telescopes. Amalthea, discovered by E. E. Barnard in 1892, is rather irregular in shape, and Voyager pictures showed craters, troughs and ridges. It was, incidentally, the last satellite to be discovered by an observer at the eye-end of a telescope – all subsequent discoveries have been made by photography or from space-craft. Metis, Adrastea, Amalthea and Thebe are closer in to Jupiter than any of the Galileans; Leda, Himalia, Lysithea, Elara, Ananke, Carme, Pasiphaë and Sinope are further out. The last four are between 13 and 15 million miles from Jupiter, and have retrograde motion, so that they are unquestionably captured asteroids. At this distance from Jupiter, their orbits are not even approximately circular.

Jupiter is much the most rewarding of the planets from the viewpoint of the amateur observer. The surface is always changing, and one never knows what will happen next. It was surely fitting that the ancients named the Giant Planet in honour of the king of the gods.

CHAPTER TWELVE

Saturn

The outermost of the planets known in ancient times was named Saturn, after Jupiter's father. It is not nearly so brilliant as Jupiter, and its yellowish hue makes it look somewhat leaden; also it moves relatively slowly against the stars, and it was generally regarded as malevolent. Yet when seen through a telescope, it is certainly the most beautiful object in the entire sky.

It is the rings which make it unique. We now know that all the giants have ring systems, but none can rival that of Saturn, and this tends to divert attention from the globe – and it is true that surface details are not striking. Basically, Saturn is not unlike Jupiter, and it too has cloud-belts and spots, but there is much less obvious activity.

During the last century it was still thought that Jupiter and Saturn must be miniature suns, and it is interesting to recall the words of R. A. Proctor in his book *Saturn and its System*, written in 1882:

> 'Over a region hundreds of thousands of square miles in extent, the glowing surface of the planet must be torn by subplanetary forces. Vast masses of intensely hot vapour must be poured forth from beneath, and rising to enormous heights, must either sweep away the enwrapping mantle of cloud which had concealed the disturbed surface, or must itself form into a mass of cloud, recognizable because of its enormous extent, and because its texture differs from that of the cloud masses surrounding it. Such a disturbance, extending in the case of Jupiter over an area as large as France, or in the case of Saturn over an area as large as Russia, would be just discernible with our most powerful telescopes. It might very well be, then, that the

> surface of either planet should appear absolutely at rest while yet disturbances of the most tremendous character were taking place in every part of the planet's globe. If over a thousand different regions, each as large as Yorkshire, the whole surface were to change from a condition of rest to such activity as corresponds with the tormented surface of seething metal, and vast clouds formed over all such regions so as to hide the actual glow of the surface, our most powerful telescopes would fail to show the faintest trace of change. And Saturn might be still more tremendously disturbed without our seeing any signs of it.'

Nothing could be further from the truth, and yet in Proctor's day the picture he painted was perfectly reasonable. Of course, Saturn is appreciably smaller than Jupiter; its equatorial diameter is 74,900 miles and the polar diameter 67,600 miles. It is also considerably further from the Sun; the mean distance is 886,000,000 miles, so that Saturn can never come much within 750,000,000 miles of the Earth. The orbital velocity is 6 miles per second, and the revolution period 29½ years. This explains why Saturn seems a slow mover in the sky. As the spin is rapid, about 10¼ hours, a Saturnian 'year' will contain 25,000 'days' – though, as with Jupiter, the rotation period is not the same all over the planet; it is quickest at the equator, slowest in the polar regions.

Saturn is much larger than any of the other planets apart from Jupiter. Its volume is over 700 times as great as that of the Earth, but its mass is only 95 times as great, because the density is so low; in fact, the overall density of the globe is less than that of water, and it has been said that if you could put Saturn into a vast ocean it would float. Though the escape velocity is high (22 miles per second) the surface gravity is not. Surface gravity depends not only upon the mass of a body, but also upon its diameter; for two globes of equal mass, the smaller – and therefore denser – will have the stronger surface pull, because an observer standing there will be closer to the centre of the globe. Not that anyone could stand on Saturn, which has a gaseous surface; but if it were possible, a man who weighs 14 stone on Earth would weigh 16 stone on Saturn. Jupiter is the only planet in the Solar System where an Earthman would feel uncomfortably heavy.

Saturn's make-up is not very different from that of Jupiter, though the core temperature is less, and is thought to be of the order of 15,000 degrees Centigrade (27,000 degrees Fahrenheit). According to the latest theoretical models, there is a rocky core rather larger than the Earth, above which comes a layer of liquid metallic hydrogen; this is overlaid by a layer of liquid molecular hydrogen, and then comes the atmosphere, with the upper clouds which we can see. The clouds contain only about 6 per cent of helium, and the rest, predictably, is mainly hydrogen. As Saturn is so much further out than Jupiter, we would expect the upper clouds to be colder, and indeed they are: around −180 degrees Centigrade, or −240 degrees Fahrenheit. Most of the ammonia has been frozen out of the upper atmosphere, and spectroscopic observations show a greater amount of methane, which does not freeze so easily.

Telescopes of fair size are needed to show much on the disk; there is a distinct resemblance to Jupiter in one of its quieter moods, but Saturn is altogether more bland. The belts appear curved; the equatorial zone is usually brightish cream in colour, and there is nothing comparable with Jupiter's Red Spot. The poles are often dusky, but there are no vivid colours anywhere.

Like Jupiter, Saturn sends out more energy than it would do if it depended solely upon the Sun, but the cause may be different, because the smaller Saturn has had ample time to lose all the heat built up during its formation. The favoured theory is that the heat is produced gravitationally by drops of liquid helium moving downward towards the core, through the less dense hydrogen. The explanation does not seem to be completely satisfactory, but as yet nobody has been able to suggest anything better.

Major outbursts are rare, but brilliant white spots near the equator are seen now and then. The first to be recorded was seen in 1876; there was another in 1903, and then, in 1933, a really spectacular outbreak. In August of that year it was discovered by a British amateur, W. T. Hay, better remembered today as Will Hay, the stage and screen comedian. It quickly became very prominent, and I remember it well; I was ten years old at the time, and very proud of the 3-inch refractor which I had managed to acquire after a long period of saving up all my pocket money and Christmas and birthday presents. (For the record, that telescope cost £7.10s; I still have it, and I still use it.) The white spot gradually lengthened, and the portion of the disk following it darkened; before long the leading edge of the spot became diffuse, while the following edge remained clear-cut. Sir Harold Spencer Jones, the Astronomer Royal, commented that the cause must be 'a mass of matter thrown up from an eruption below the visible surface, encountering a current travelling with greater speed than the erupted matter, which was carried forward with the current while still being fed from the following end'. The spot soon faded, and in a few months it had become nothing more than a bright zone stretching right round the planet; then it disappeared completely.

The next white spot, that of 1960, was much less striking, but observers of the planet had a treat in 1990. On 25 September an American amateur, Stuart Wilber, detected a brilliant white spot in much the same latitude as the earlier ones. It was soon confirmed – and indeed it could hardly be missed. I had been at a conference abroad, but as soon as I returned, a week later, I turned my 15-inch reflector towards Saturn, and the white spot was glaringly obvious. The sequence of events followed the usual pattern. Within a few days the spot had been spread out by the strong equatorial winds, and had become a 'cloud' over 9000 miles long; by mid-October it had been transformed into a bright zone all round the equator. The brilliance gradually faded, and in a few months everything was normal once more.

There is an interesting periodicity here. The recorded white spots have been in 1876, 1903, 1933, 1960 and 1990; the intervals between outbreaks have been 27 years, 30 years, 27 years and 30 years. This is very close to Saturn's orbital period of 29½ years, and although I am always wary of coincidence-hunting, it does look as though there may be a connection here. Observers will be on the watch for a new white spot around 2020. I am afraid that I am unlikely to see it, since I will then have reached the advanced age of ninety-seven, but no doubt other observers will be more fortunate. The spots are important because they tell us a good deal about conditions below the visible surface, and also help in measuring rotation periods.

Therefore, it is always wise to scan the globe closely, and avoid being too preoccupied with the glorious rings. It is in making observations of this sort that the well-equipped amateur really comes into his own.

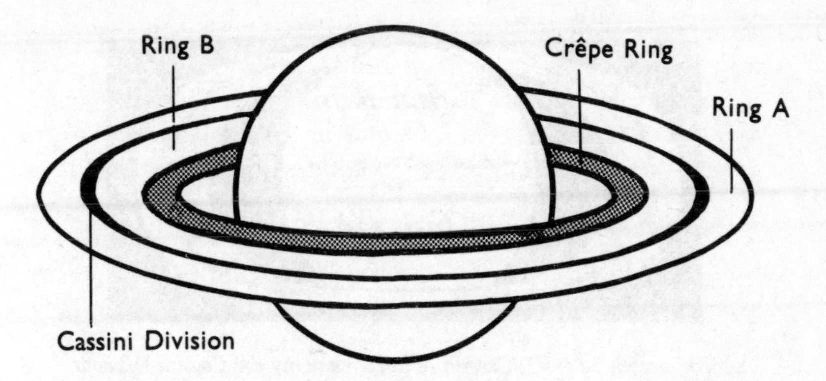

The principal rings of Saturn, A and B, separated by the Cassini Division, and C, the Crêpe Ring.

But, of course, it is the ring system which makes Saturn so magnificent. A small telescope will show it, and the rings have been known ever since the seventeenth century; Galileo saw them, though not clearly enough for him to realize what they were. He mistook Saturn for a triple planet, and was puzzled to find that after a few years the strange aspect vanished, leaving Saturn normal in shape. We have found out the answer to the problem, but Galileo never did. Soon after he began observing, the ring system became edgewise-on to the Earth, and Galileo's low-powered telescope could no longer show it at all.

In 1659 Christiaan Huygens, who was probably the best observer of his time, issued a famous anagram* in which he announced that Saturn is surrounded by 'a flat ring, which nowhere touches the body of the planet, and is inclined to the ecliptic'. He was correct so far as he went, but his theory met with a surprising amount of opposition; for example a French Jesuit mathematician, Honoré Fabri, claimed that the odd aspect of Saturn was due to the presence of four satellites – two dark and close in, the other two bright and further out. Not for some years

* The anagram read: aaaaaaa ccccc d eeeee g h iiiiiii llll mm nnnnnnnnn oooo pp q rr s ttttt uuuuu. Rearrange these letters in their proper order, and you will have the Latin sentence: *Annulo cingitur, tenui, plano, nusquam cohærente, ad eclipticam inclinato.* In those days it was quite common for discoveries to be announced in anagram form, to establish priority. Galileo himself had done so when he first detected the phases of Venus.

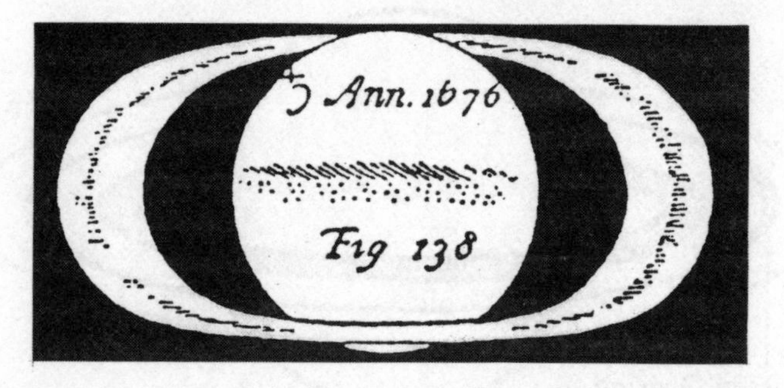

Saturn, drawn by G. D. Cassini in 1676, showing the Cassini Division for the first time.

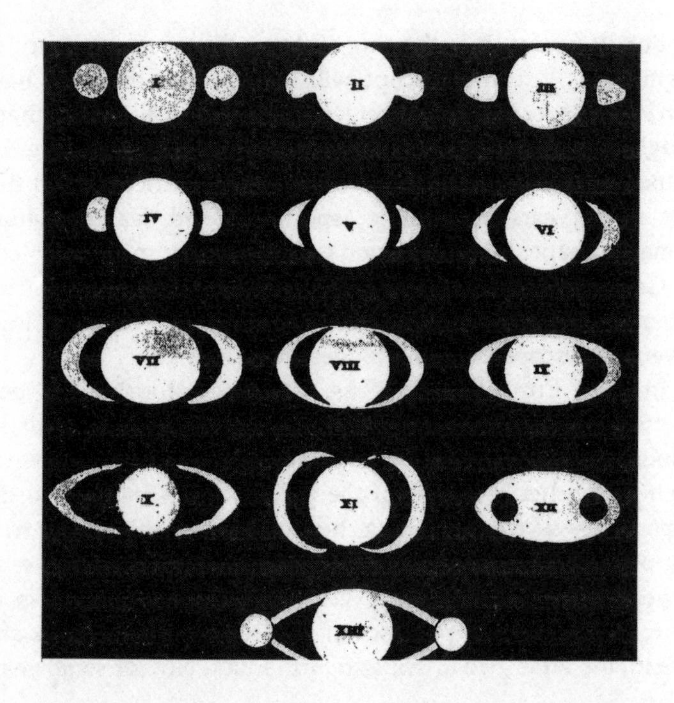

Early drawings of Saturn. I Galileo, 1610. II Scheiner, 1614. III Riccioli, 1643. IV–VII Hevelius, c.1645. VIII–IX Riccioli, 1648–50. X Divini, 1646–8. XI Fontana, 1636. XIII Fontana, 1644–5. These were all made before Huygens discovered that the cause of Saturn's unusual appearance is due to a ring.

was Huygens' ring accepted by all astronomers. Some of the drawings made at this period look very peculiar, but this is hardly surprising in view of the weak telescopes which had to be used.

There are three principal rings, two bright and one dusky. The outer bright ring (A) is 9000 miles wide; then comes a well-marked gap, discovered by G. D. Cassini in 1675 and therefore called Cassini's Division. It is 2500 miles wide, and separates Ring A from the brightest ring, B, which has a width of 16,000 miles. The ring described by Huygens was a combination of A and B.

Rings A and B are not alike. B is much the brighter of the two, and is less transparent; even a small telescope of good quality will show the difference, and when the ring system is suitably tilted I have never had any problems in seeing the Cassini Division with my 3-inch refractor. A narrow division in Ring A, discovered by J. F. Encke and named after him, is much more difficult, and is hard to see when the rings are almost edgewise-on to us – as they will be in the mid-1990s.

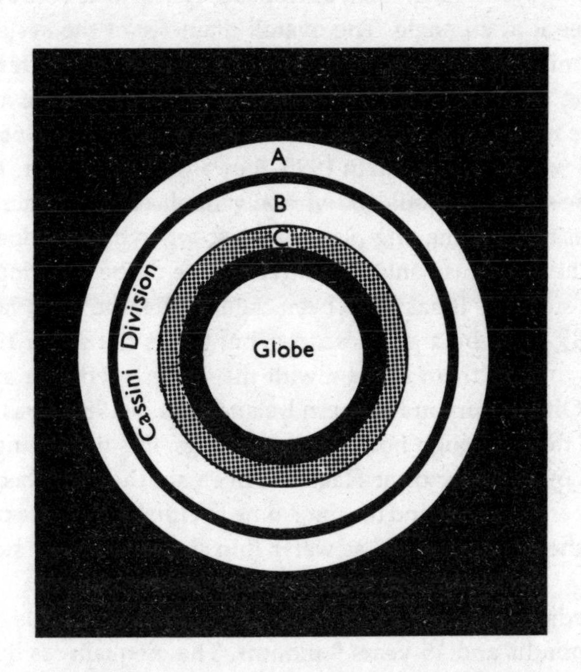

Bird's-eye view of the principal rings (A,B,C).

Between Ring B and the planet is a third ring, Ring C, often called the Crêpe or Dusky Ring. It was first recognized in 1850 by two independent observers, W. Bond in America and W. R. Dawes in England. It is by no means conspicuous, and is semi-transparent. The width is about 12,000 miles.

Well before the Space Age there were reports of other, less evident rings. One of these was said to be closer in than the Crêpe Ring, and was referred to as Ring D; another, outside the main system, was reported by the French astronomer G. Fournier in 1907, and, confusingly, was also referred to as Ring D. I was always rather sceptical, because I looked for these rings, using very powerful telescopes (including the Meudon 33-inch refractor) with a total lack of success. The answers came much later, with the flights of the Pioneer and Voyager spacecraft.

Shadow effects are fascinating. The globe casts shadow on the rings, giving the false impression of a break; the shadows cast by the rings on the disk of Saturn are easy to see, and unwary observers have often mistaken them for belts.

The ring system is circular, but looks elliptical to us because we always see it at an angle. The overall diameter of the system is 169,000 miles, but the rings are remarkably thin, and this has interesting consequences from the observational point of view. When the rings are edgewise-on to us, as happened in 1966 and 1980 and will happen again in 1995, they almost disappear. To be more precise, they should be virtually unobservable when the Earth goes through the ring-plane and also when the Sun does so, since in the latter case only the extreme edge of the thin ring can catch the sunlight. It has often been claimed that the rings vanish completely, even in large telescopes, but this is not so. In 1966 I was able to keep them in view with the 10-inch refractor at the Armagh Observatory in Northern Ireland, and in 1980 I was able to follow them through both plane crossings, this time using the 24-inch Lowell refractor at Flagstaff. Even so, the rings become very faint and elusive, and for some time to either side of the plane crossing they show up only as wafer-thin streaks to either side of the disk.

The edgewise presentation occurs at alternate intervals of 13 years 9 months and 15 years 9 months. The inequality is due to Saturn's orbital eccentricity. During the shorter interval the south

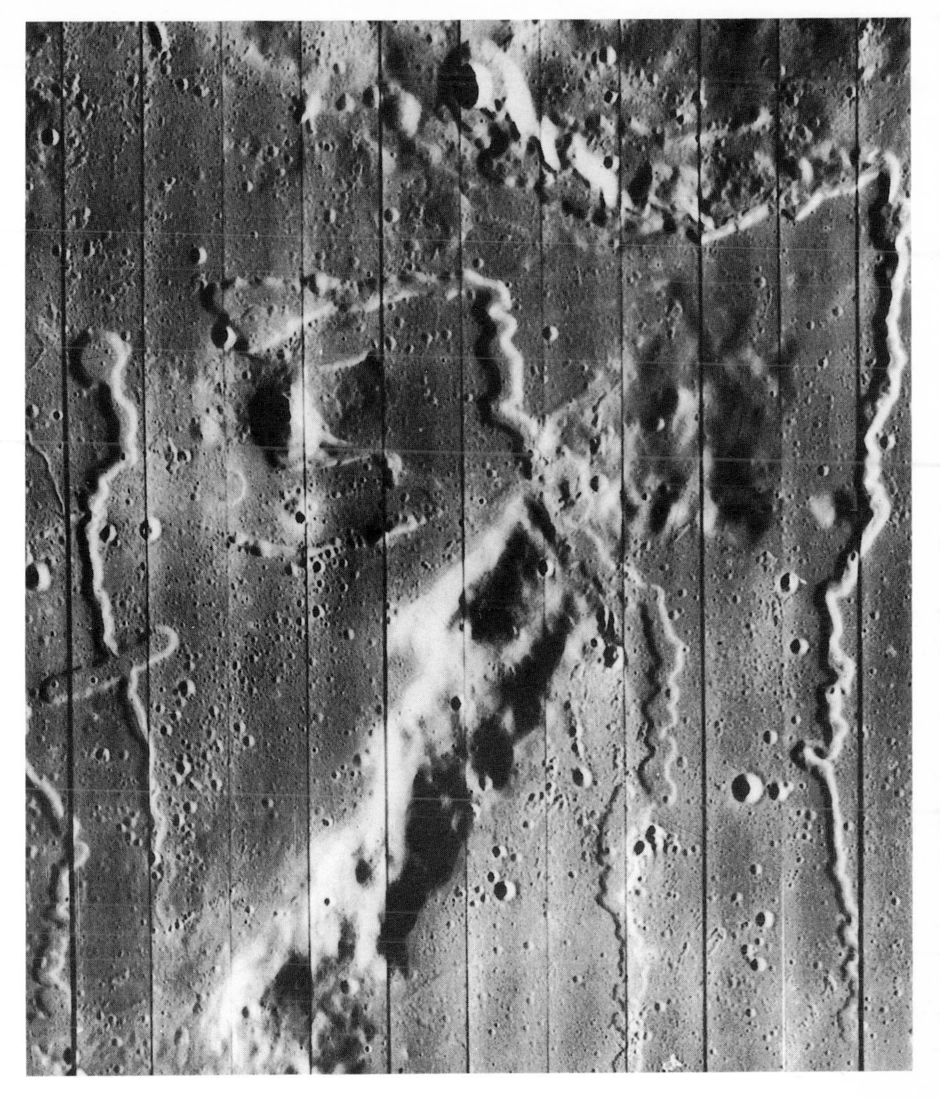

Mars, from Viking. It is difficult to resist the conclusion that these features are old riverbeds cut by running water.

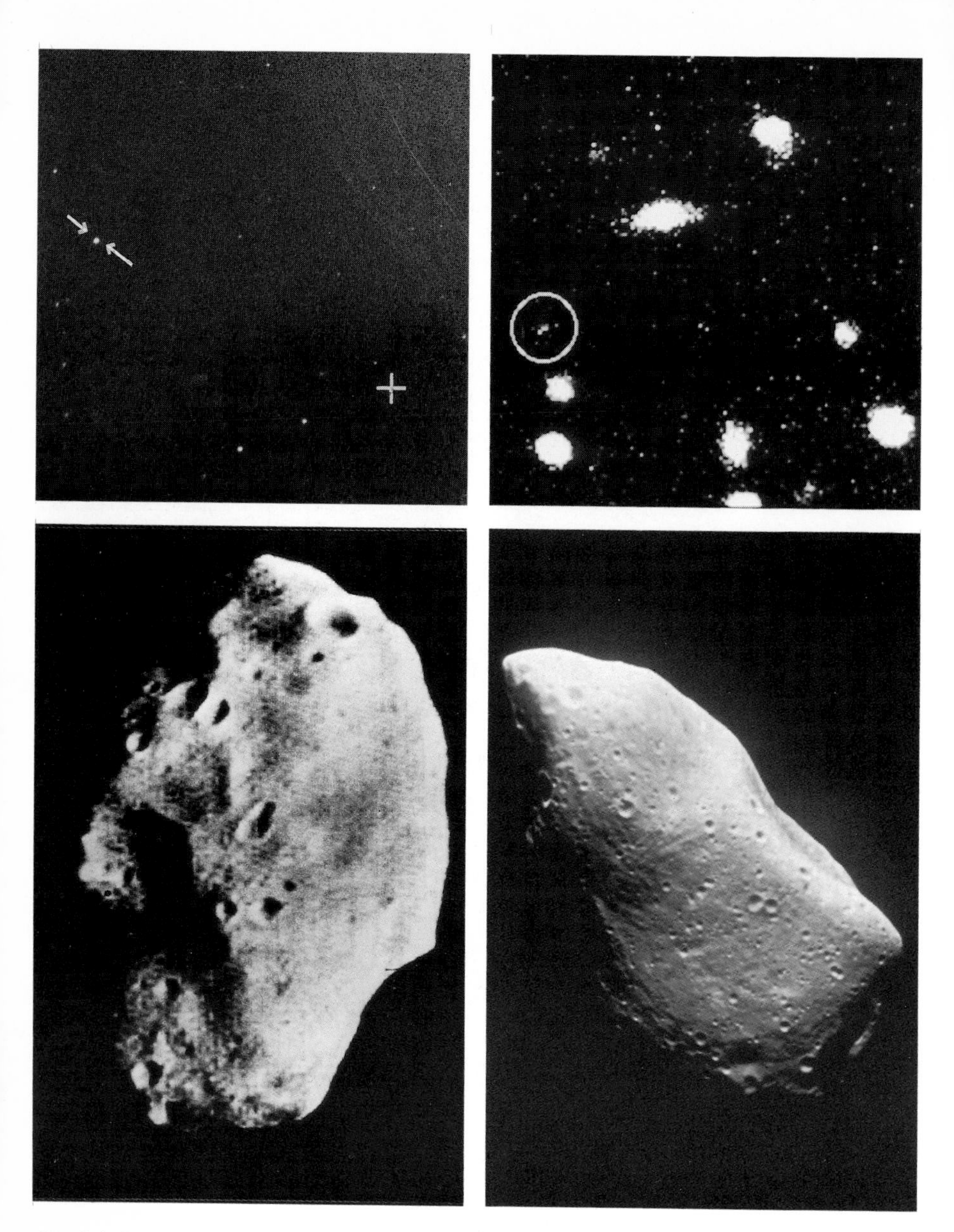

Top Left: Vesta, the brightest asteroid (photograph by F. Acfield, 6-in, OG). The cross indicates Vesta's position on the previous night.

Top Right: The most remote asteroid: 1992 QB1 (circled) 28 September 1992 (A. Smette and C. Vanderreist, New Technology Telescope, European Southern Observatory, La Silla, Chile). The field is 3.3 arc minutes across; QB1 is of magnitude 23. Reproduced by kind permission of the ESO.

Bottom Left: Phobos, the senior satellite of Mars, from Mariner 9. It is probably a captured asteroid.

Bottom Right: Gaspra, photographed in 1991 by the Galileo spacecraft. It is approximately the same size as Phobos.

CHANGING JUPITER

Above Left: Photograph: H. E. Dall, 14 October 1975 (15½-in refl.).

Above Right: Drawing: Patrick Moore, 31 December 1987, 17.50 (15-in refl. × 400).

Middle: Jupiter, from Voyager 1. Two satellites are shown: Io (against the Red Spot) and Europa.

Bottom: Details in the Red Spot, from Voyager 1.

Right: Saturn, 29 March 1980 (Patrick Moore, 15-in refl. × 400, 19.45). Rings edgewise-on (this was the last edgewise presentation to date).

Middle Right: Saturn, 2 June 1983 (Paul Doherty, 15-in refl. × 360. 21.36). The rings are opening.

Below: Saturn from the Hubble Space Telescope, 1990. The ring system is now wide open as seen from Earth.

Top: Saturn from 47,000,000 miles: Voyager 1, 17 September 1980. Several satellites are shown: Titan (*Upper Right*), Dione, Tethys and Rhea (*Upper Left*), Mimas and Enceladus (*Lower Right*).

Bottom: The white spot on Saturn: Hubble Space Telescope, 9 November 1990. The pictures were taken with the Planetary Camera. By then the 'spot' had really become a bright zone, spreading round the planet. The sharpness of the HST image reveals a very turbulent atmosphere, similar to the cloud system trailing the Great Red Spot on Jupiter.

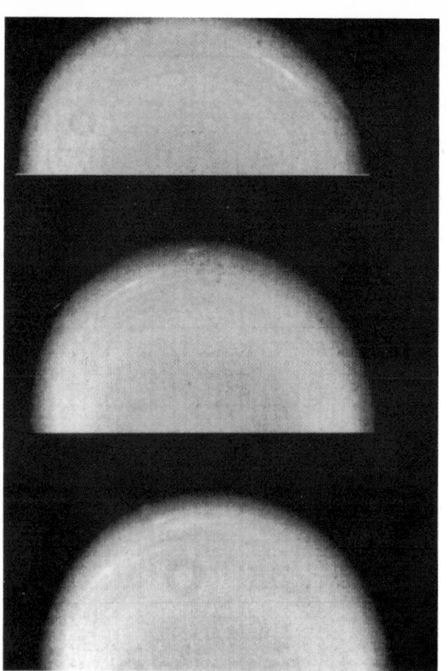

Left: Uranus from Voyager 2, showing small, streaky clouds, first detected on 14 January 1986 from 8,000,000 miles. The movement of the clouds between the exposures is evident – remember that the pole lies near the centre of the disk!

Below: The rings of Uranus, from Voyager 2; range 147,000 miles. The resolution is 20 miles. There is a great deal of 'dust' in the ring-system. The long exposure produced star trails as well as a noticeable, non-uniform smear.

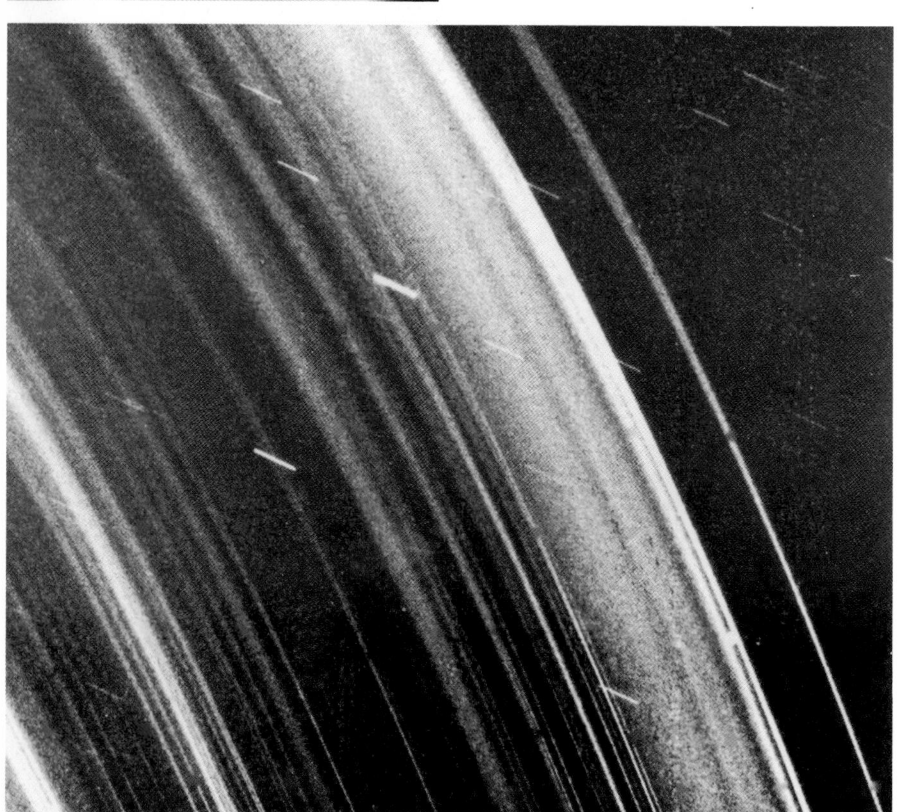

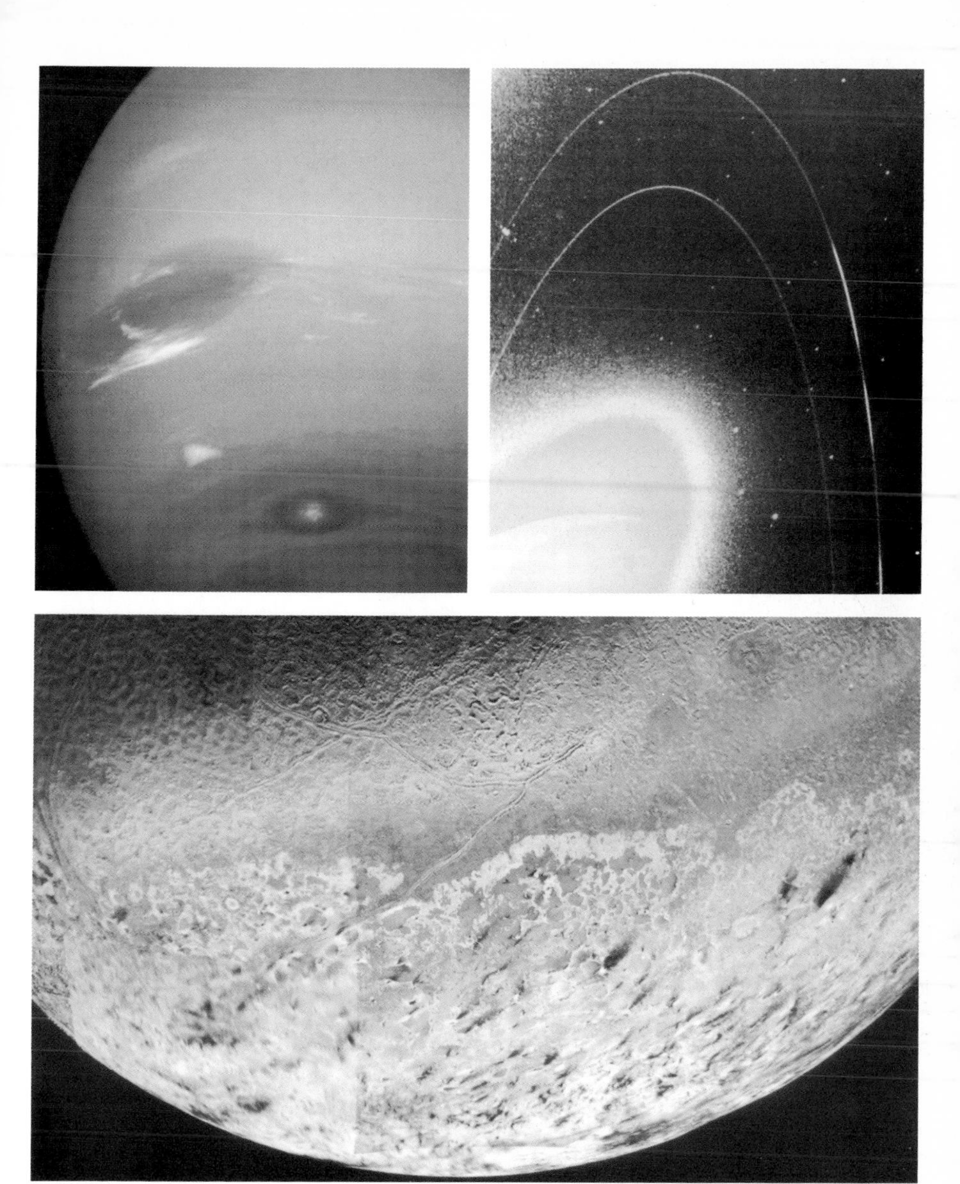

Top Left: The blue planet Neptune, from Voyager 2. At the north (*Top*) is the Great Dark Spot; to the south of it is the Scooter, and still further south is the second dark spot, D2. Each feature moves eastward at different velocity, so it is not often that they appear close to one another, as in this picture.

Top Right: The rings of Neptune, from Voyager 2 backlit by the Sun as Voyager swept by Neptune. The image of the planet was overexposed to make it possible to show ring details. In this view the rings appear bright, as tiny ring particles scatter sunlight towards Voyager's camera. Note the 'arcs', now known to be parts of a continuous ring. The picture was taken from a range of 680,000 miles.

Bottom: Triton, from Voyager 2. The polar region (Uhlanga Regio) is covered with pink nitrogen 'snow'; the dark streaks are the geysers.

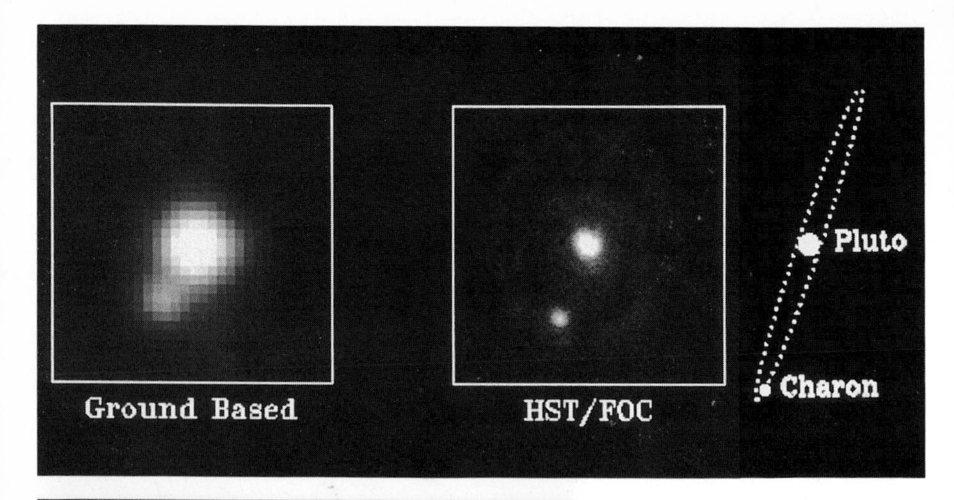

Above: Pluto and Charon. (*Top Left*): The best view from ground-based equipment. (*Top Right*): The Hubble Space Telescope picture. (*Below*): the apparent movements of Pluto and Charon at this time – the late 1980s. At maximum separation, the distance between Pluto and Charon is no more than 0.9 of an arc-second.

Left: Last view of Neptune and Triton from Voyager 2; the spacecraft was then 3,000,000 miles from Neptune. Both planet and satellite show up as crescents.

pole of Saturn is tilted sunward, so that part of the northern hemisphere is hidden by the rings; Saturn passes through perihelion, and so is moving at its fastest. During the longer interval the north pole is turned sunward, so that parts of the southern hemisphere are covered up; Saturn passes through aphelion, and is moving more slowly. At the moment (1993) it is the southern hemisphere which is partly covered; the next edgewise presentation is due in 1995. The rings lie exactly in the plane of Saturn's equator, which is tilted to the orbital plane by just over 26½ degrees.

The two main rings, A and B, look so solid that it was natural for early telescopic workers to class them as rigid sheets. Not everyone agreed; in 1705 J. Cassini proposed that the rings were made up of small particles revolving round Saturn – and this was a shrewd guess, though it was not until the nineteenth century that it was confirmed.

In 1848 Edouard Roche, of France, proved that if a solid body with virtually no gravitational cohesion comes too close to a planet (or other object) it will be broken up. The edge of the danger-zone is known as the Roche limit; it depends on the size and mass of the planet concerned. The rings lie well inside the Roche limit for Saturn,

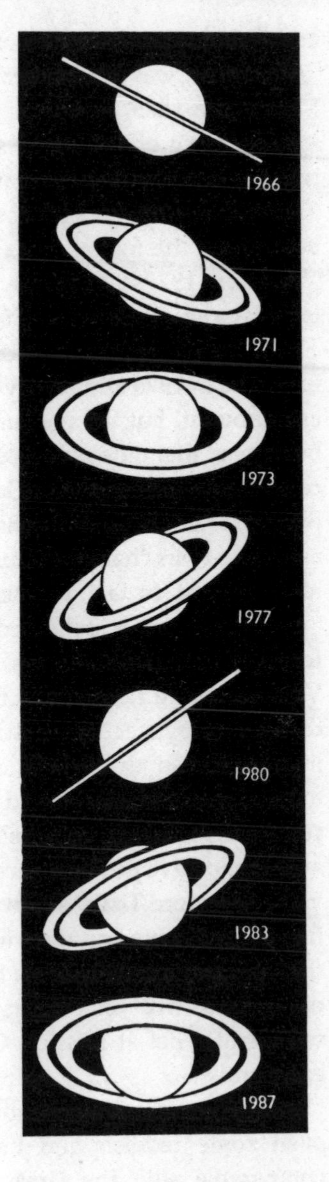

The last cycle of Saturn's rings, from 1966 to 1980. The rings were wide open in 1987, and are now closing, so that they will again be edgewise-on in 1995.

and therefore they would soon be broken up if they were solid or liquid. This was demonstrated mathematically by James Clerk Maxwell in 1875, and then, twenty years later, J. E. Keeler made spectroscopic observations showing that the inner parts of the rings revolve round Saturn at a greater speed than the outer parts. This, of course, is in agreement with Kepler's Laws. Each particle behaves in the manner of a tiny, independent moonlet.

Before 1979 it was tacitly assumed that the rings were more or less flat and regular, but the space missions showed otherwise. The first foray came with Pioneer 11, which, as we have seen, was launched in 1973 and surveyed Jupiter. Saturn was more or less an afterthought, but the encounter was very useful indeed, mainly because at that time nobody knew the extent of the danger due to collisions with débris near Saturn. Pioneer was scheduled to pass within 13,000 miles of the cloud-tops, and actually did so; estimates of its chances of survival ranged from 99 per cent down to 1 per cent. In fact, it emerged unscathed.

The Voyagers followed in 1980 and 1981 respectively, following their encounters with Jupiter in 1979. Though the Voyagers were near-twins, their rôles after leaving Jupiter were different. Voyager 1 was scheduled to survey not only Saturn itself but also the largest satellite, Titan, which was known to have an atmosphere and to be of exceptional interest; to do so it would have to move well away from the plane of the eclipitic, and would have no more encounters. In this case Voyager 2 would virtually ignore Titan, and would be able to go on to rendezvous first with Uranus and then with Neptune. Had Voyager 1 failed, then Voyager 2 would have had to survey Titan, thereby missing out on the two remote giants. It is not surprising that there was great relief at Mission Control when Voyager 1 performed excellently.

Superb pictures were obtained of Saturn's globe; there were even some reddish and brownish spots, though in no way comparable with the Great Red Spot on Jupiter. Windspeeds reached up to 900 mph, faster even than those on Jupiter, and were symmetrical with the equator. The magnetic field was twenty times weaker than Jupiter's, but still a thousand times stronger than that of the Earth. It was found that the magnetic axis coincided with the axis of rotation, so that on Saturn a magnetic compass would indeed point to the pole. Auroræ were recorded, but, predictably, were much weaker than those of Jupiter.

As Voyager 1 drew in towards Saturn, it was clear that the rings were much more complicated than anyone had expected. They were divided into thousands of narrow ringlets and minor gaps, and the whole aspect was extraordinary by any standards. It had always been believed that the well-known divisions (Cassini's and Encke's) were due to the gravitational effects of the satellites, particularly the innermost one known before the Voyager missions (Mimas). This was acceptable when only a few very well-marked divisions were known, but the complexity of the system meant that satellite perturbations could not be the complete answer, and even now our theories about the dynamics of Saturn's ring system remain rather vague.

The Cassini Division was not empty; it contained ringlets as well as general débris, while Ring B showed strange radial spokes, 6000 miles long, which persisted for some hours after emerging from the shadow of the globe. They ought not to have been able to form; remember that according to Kepler's Laws, a particle moving in the inner region will have a greater speed than a particle further out, so that no radial feature should have been possible. Yet they were there, and very distinct. My immediate reaction was that the spokes were due to particles elevated away from the ring-plane by magnetic forces, and then swept along by the magnetic field lines. This does seem to be the most likely explanation. Checking on some old drawings made by Earth-based observers showed that the spokes had been recorded now and then, notably by Antoniadi.

New rings were found. The so-called Ring D, extending down to the cloud-tops, was not really a ring at all, but merely a region of diffuse material. (I had always been suspicious of some of the earlier published accounts of it, which I put down to observers 'seeing' what they hoped to see.) However, there was a new ring outside Ring A, presumably the same as Fournier's; it is officially called Ring F, and is strangely complex, with interwoven strands. Ring G is very tenuous, and occupies most of the space between the orbit of Mimas, innermost of the larger satellites, and the two dwarf moons, Janus and Epimetheus, which share the same path. Finally there is Ring E, which is even more tenuous, and is at its brightest just inside the orbit of the second major satellite, Enceladus.

Not even Voyager could show the ring particles separately, but it seems that the particle sizes range from pebbles up to blocks of ice several feet across. There is also a rarefied cloud of hydrogen extending to about 40,000 miles above and below the ring-plane. As for the composition – well, it seems definite that the ring particles are made up of ordinary water ice.

Before the Space Age, Saturn was known to have nine satellites. The family is very different from that of Jupiter. Instead of four large satellites and a dozen small ones, Saturn has one major satellite (Titan) and several of medium size. Rhea and Iapetus are around 900 miles in diameter; Dione and Tethys about 700; Mimas, Enceladus and Hyperion between 170 and just over 300. The other known satellite was Phœbe, a mere 140 miles across, moving round Saturn at 8,000,000 miles in a retrograde sense, so that clearly it qualifies as an ex-asteroid. Since then nine more satellites have been found. Pan, Atlas, Prometheus, Pandora, Epimetheus and Janus are closer in than Mimas; Telesto and Calypso move in the same orbit as Tethys, while Dione has one 'Trojan', Helene. No doubt other small satellites exist, and the grand total is probably well over twenty. All the new satellites are very small, and only Epimetheus has a diameter of as much as 100 miles.

The latest satellite to be identified, Pan, actually moves inside the Encke Division in Ring A. Prometheus and Pandora are known as 'shepherd' satellites, because they move to either side of Ring F and keep it stable. Prometheus lies slightly outside the ring, and is therefore moving more slowly than the ring particles; if a particle strays, Prometheus will slow it down and make it drop back to a lower orbit. If a particle strays inward towards Pandora, closer in to Saturn, it will be speeded up and returned to the main ring. Janus and Epimetheus are certainly fragments of an originally larger single body; they have almost identical periods, and every four years they approach each other, so that mutual interactions make them exchange orbits – rather in the manner of two moons playing a game of cosmic musical chairs! Most of these dwarf satellites are very irregular in shape.

Either Janus or Epimetheus seems to have been seen in 1966

by Audouin Dollfus, using the powerful refractor at the Pic du Midi Observatory in the Pyrenees. This was the year of the edgewise presentation of the rings, providing an excellent opportunity for seeking close-in satellites. I have to confess that I missed a golden opportunity. I was making a series of observations of the known satellites with the 10-inch refractor at Armagh, but having no tables to hand I did not at once reduce the observations systematically; I was waiting until the series had been completed, which would have been in the following January. After Dollfus' announcement I did some checking, and found that I had recorded Janus several times between July and November without realizing that it was new. Of course I can claim absolutely no credit; it was a good example of overlooking the unexpected. The sole value of my observations was that they showed Janus to be slightly brighter than was at first thought.

Titan is the largest of the satellites, and is indeed the largest satellite in the Solar System apart from Ganymede. It is bright enough to be seen with a small telescope, and I have been told that it can even be glimpsed with binoculars, though I have never managed to do so myself. In 1944 it was found to have an atmosphere, and pre-Voyager it was generally thought that this atmosphere was likely to be made up chiefly of methane.

Rhea can be seen easily with a 3-inch refractor, and Dione and Tethys with more difficulty. Iapetus is curious inasmuch as it is much brighter when west of Saturn than when to the east of the planet; at its best it can equal Rhea, but at its faintest it is beyond the range of a 3-inch. This odd behaviour had been known ever since the satellite had been discovered, by G. D. Cassini in 1671, and there seemed only one plausible way to account for it. Like most other major planetary satellites, Iapetus has captured rotation; that is to say, it keeps the same face turned towards its primary all the time, and the orbital period is the same as the rotation period, 79 days in the case of Iapetus. This means that at western elongations the more reflective hemisphere is always facing us.

Of the other satellites, I can see Mimas and Enceladus with my 12½-inch reflector, and I have glimpsed Hyperion, which is not too difficult with my 15-inch telescope; it is best sought when in conjunction with Titan. Phœbe is much fainter, and the recently-discovered satellites are beyond the range of amateur telescopes under normal conditions. My 1966 sightings of Janus were obtained when the situation was more or less ideal.

Titan, a prime target for Voyager 1, provided plenty of shocks. There was indeed a thick atmosphere, hiding the surface completely, but it was found to be made up chiefly of nitrogen, plus a good deal of methane. The ground pressure is more than one and a half times the pressure of the Earth's air at sea-level. Voyager 1 flew past at only 4030 miles, but showed nothing except the top of a layer whch might be called orange smog. The surface temperature has been given as −290 degrees Fahrenheit (−180 degrees Centigrade), and this is significant, because it means that on Titan methane can exist as a solid, a liquid or a gas – just as H_2O can do on Earth, as ice, liquid water or water vapour. There may be seas on Titan, but they will not be like our seas, and are likely to be made up of some sort of chemical substance. Ethane and methane are possibilities.

Titan seems to be far too cold for life to have appeared there, but there is plenty of organic matter – hence the orange hue – and the satellite does seem to have all the ingredients for life. We may know more in 2004, because a new probe is scheduled to go there and make a controlled landing.

One other point is worth making. Titan's escape velocity is about the same as that of our airless Moon, but Titan has been able to retain its atmosphere because it is much colder – and lowering the temperature means that atoms and molecules move around more slowly, so that they find it less easy to escape. It has been suggested that when the Sun becomes more luminous, several thousands of millions of years hence, Titan will become warm enough to support life. Unfortunately, however, the increased heat will make Titan lose all its atmosphere very quickly.

Between them the Voyagers obtained good images of all the larger satellites apart from Phœbe, which was badly placed. Mimas is icy and cratered, with one huge crater, now named Herschel, which has one-third the diameter of Mimas itself; Enceladus is icy and smooth, with small craterlets; Tethys is almost pure ice, with one tremendous trench running for more than half-way round the globe. Dione, only slightly larger than Tethys but considerably more massive, has hemispheres of unequal brightness, with bright, wispy features and some large craters. Rhea's surface gives every impression of being very ancient, and is almost as heavily cratered as Callisto in Jupiter's system. Hyperion is an oddity; it is irregular in form, measuring 224 × 174 × 140 miles, and is shaped rather like a hamburger. It takes 21.3 days to go once round Saturn, but its rotation is not captured; Hyperion seems to be tumbling in its orbit, and the axial spin has been described as chaotic. It has been suggested that Hyperion may be a fragment of a former larger body, but there is no sign of the other half.

Iapetus did indeed prove to have a bright following hemisphere, as reflective as snow, and a leading hemisphere which is as black as a blackboard. Theorists were faced with what I have called the Zebra Problem: is a zebra a white animal with black stripes, or a black animal with white stripes? In the case of Iapetus we can find out. Its movements, and its effects on other satellites, show that its density is not much greater than that of water, so that the globe is made up largely of ice. The dark stain is more of a puzzle. There were suggestions that it might be due to 'dust' wafted on to Iapetus from the outermost satellite, Phœbe, which – as far as we can tell from the one unsatisfactory Voyager picture – is darkish, and unlike the icy satellites. However, Iapetus and Phœbe never come within 6 million miles of each other, and in any case the Iapetus stain is of the wrong colour. Either the satellite was hit by a comet in the remote past, or else (more probably) the dark matter has welled up from below the icy crust.

Eclipses, transits and shadow transits of the satellites can be observed, though much less easily than with the Galilean satellites of Jupiter. Only for Titan are the phenomena within the range of a small telescope, and this is a pity, because the orbits of the smaller satellites are not known with absolute precision, and timings of the eclipses and transits would be useful. Mutual phenomena have been seen now and then; for instance, on 8 April 1921 A. E. Levin and L. J. Comrie observed an eclipse of Rhea by the shadow of Titan.

Pickering's discovery of Phœbe was made with a 24-inch telescope at Arequipa in Perú, the southern station of the Harvard College Observatory. Six years later he announced that he had found a new satellite moving in an orbit between those of Rhea and Titan. He named it Themis, and for some time its existence was regarded as well established, but it has never been seen since, and probably does not exist.

One day, perhaps, men will stand upon some of Saturn's satellites and admire the view. The most spectacular scenes will be obtained from Phœbe and Iapetus, since only these two satellites have orbits which are appreciably inclined to the plane of the rings; the others move practically in the ring-plane, which also coincides with Saturn's equator. Go to Dione, for instance, and the rings will be permanently edge-on. Travel to the Saturnian system lies in the far future, but at least we can enjoy the spectacle as seen through our telescopes. In my view, at least, Saturn is the supreme gem of the heavens.

CHAPTER THIRTEEN

Uranus

Five planets were known in ancient times. Together with the Sun and Moon, this gave seven members of the Solar System, and seven was the 'mystical' number; nothing could be tidier – and little thought was given to the possibility of a new major planet. There matters rested until 1781, when a discovery by a then-unknown musician-astronomer took the astronomical world by surprise.

William Herschel was born in Hanover, but came to England when still a young man, and became a professional organist; he settled in the fashionable spa of Bath, and soon built up an excellent reputation. His hobby was astronomy, and he began to make reflecting telescopes. He was also a first-class observer, and on the night of 13 March 1781 he was using one of his home-made telescopes to examine stars in the constellation of Gemini, the Twins, when he found something which was to alter his whole life. To quote from his own account:

'In examining the small stars in the neighbourhood of H Geminorum I perceived one that appeared visibly larger than the rest; being struck with its uncommon appearance I compared it to H Geminorum and the small star in the quartile between Auriga and Gemini, and finding it so much larger than either of them, I suspected it to be a comet.'

Comets are interesting, but by no means uncommon, and at the time Herschel was not particularly excited. His first paper on the subject was headed 'An Account of a Comet', and he had no idea of the importance of what he had found. Then the mathematicians set to work, and the orbit was worked out. It was not in the least like that of a comet. Instead, the object was a planet, much more remote than Saturn, and moving round the Sun at a distance of 1,782,000,000 miles in a period of 84 years.

Herschel suggested the name of Georgium Sidus, or the Georgian Star, in honour of King George III of England and Hanover, who gave Herschel the status of King's Astronomer (not Astronomer Royal; that position was filled by Nevil Maskelyne) and gave him a pension which enabled him to give up music as a profession and devote all his time to astronomy. Foreign astronomers were not impressed with the name, and neither did they accept 'Herschel' in honour of the discoverer.* Finally the mythological system prevailed, and the new planet became Uranus, after the original ruler of Olympus.

It is often said that the discovery was due to sheer chance, but this is unfair to Herschel, who was busy upon a systematic 'review of the heavens'. As he said in a letter written to a friend of his, Dr Hutton: 'Had business prevented me that evening, I must have found it the next, and the goodness of my telescope was such that I must have perceived its visible planetary disk as soon as I looked at it.' Probably the new planet could hardly have been missed for much longer even if Herschel had never made a telescope; it would surely have been found by Schröter's 'celestial police' in the first years of the new century, when they set out to hunt for the missing planet between the orbits of Mars and Jupiter.

Though Herschel was the first to recognize Uranus, he was not the first to see it. It had been recorded several times in earlier years, and the first Astronomer Royal, John Flamsteed, saw it half a dozen times between 1690 and 1715 without realizing that it was anything but an ordinary star; he even gave it a stellar designation – 34 Tauri. Keen-sighted people can see it easily enough with the naked eye when they know where to look; the mean magnitude is 5.7.

Uranus ranks as a giant. Though it is much smaller than Jupiter or Saturn it is far larger than the Earth, with an equatorial diameter of 31,770 miles; the polar diameter is rather less, because the globe is appreciably flattened. Uranus is denser than water, and much denser than Saturn. Although the volume is 67 times that of the Earth, the mass is only 14½ times as great; the escape velocity is 14

* A telling point was made much later by a well-known amateur astronomer, Admiral W. H. Smyth. The name 'Herschel' might be acceptable, as would 'Le Verrier' for the co-discoverer of Neptune in 1846; but suppose either planet had been discovered by contemporary astronomers who rejoiced in the names of Bugge and Funk?

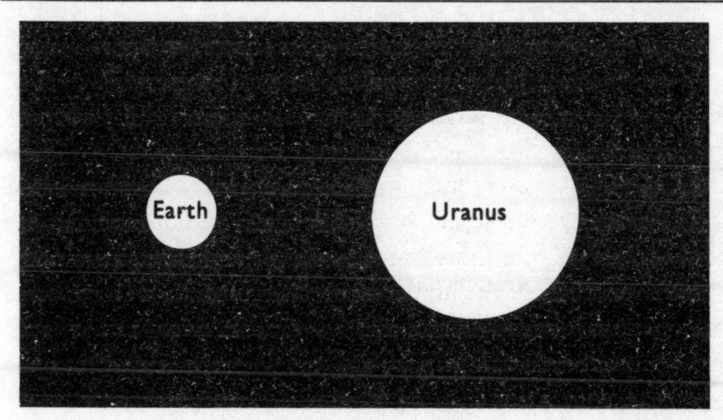

Uranus and Earth compared.

miles per second, with a surface gravity only very slightly greater than that of the Earth.

Seen through a telescope, Uranus shows a pale bluish-green disk. There is no mystery about this. The cloud-tops are so cold that the methane freezes out into clouds which cover the ammonia clouds lower down. Methane absorbs light of long wavelength (i.e. red and orange) but not blue and green; hence Uranus' distinctive colour. In the atmosphere hydrogen is dominant, with about 15 per cent of helium.

It would be quite wrong to think of Uranus as merely a smaller version of Jupiter or Saturn. The internal make-up is quite different. According to the latest models there is a core – which may or may not be well-defined – surrounded by dense layers in which gases are mixed with 'ices', i.e. substances which would be frozen out if they were as cold as the upper clouds. It seems that the bulk of this mixture is composed of water, and that water, ammonia and methane condense in that order to produce the thick, icy cloud-layers.

This sort of situation was expected long before the Voyager 2 mission, and it has been confirmed from the spacecraft results. It is fair to say that Uranus and the outer giant Neptune are near-twins, and that the Jupiter/Saturn pair is strikingly different from the Uranus/Neptune pair. Yet even between the outermost giants there are differences; Uranus is unusual in having no internal heat-source, or at best a very weak one, and it also has a most remarkable axial tilt.

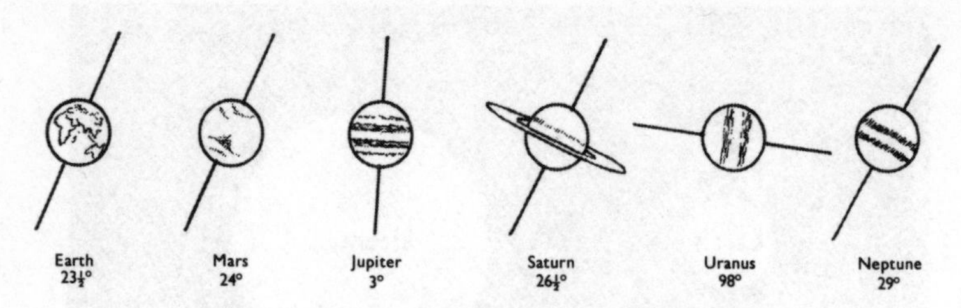

Axial inclinations of the Planets.

It is a quick spinner, though not so rapid as Jupiter or Saturn. We now know that the rotation period is 17.24 hours (rather longer than had been thought before the flight of Voyager 2), but as seen from Earth the pole may sometimes lie in the centre of the disk.

Most of the planets have their rotational axes moderately inclined to the perpendiculars to their orbits. With the Earth the angle is 23½ degrees; Mars is much the same; Saturn and Neptune slightly more, while Jupiter and Mercury are almost 'upright'. Uranus has its own way of behaving. The axial tilt is 98 degrees, which is more than a right angle, so that the rotation is technically retrograde. This means that the Uranian seasons are peculiar, to put it mildly. First one pole and then the other will be in darkness for 21 Earth years at a time, with a corresponding midnight sun at the opposite pole, though for the rest of the revolution period the conditions will be less extreme.

But which is the 'north' pole, and which is the 'south'? This is a question which is not nearly so easy to answer as might be thought – and I well remember that at the Press briefings at Mission Control in Pasadena, during the Voyager 2 encounter in 1986, there was endless discussion about it. The International Astronomical Union has decreed that all poles above the ecliptic (that is to say, the plane of the Earth's orbit) are north poles, and all those below the ecliptic are south poles; in this case it was Uranus' south pole which was in sunlight when Voyager 2 passed by. However, the Voyager team members reversed this, and referred to the sunlit pole as the north pole. Take your pick; I propose to follow the IAU decision.

The sharp tilt means that from Earth we sometimes look

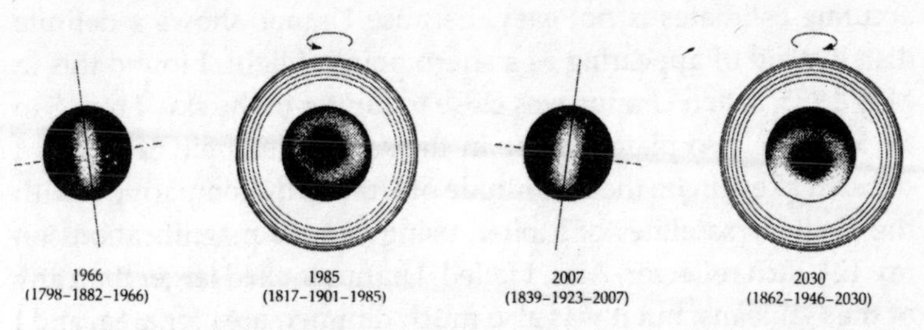

The changing presentation of Uranus. Equator-on 1966 and 2007; pole-on in 1985 and 2030.

straight at a pole and sometimes at the equator. In 1946 the north pole lay in the centre of the disk, with the equator running round the edge; by 1966 the equator ran straight 'up and down', with the poles lying at opposite limbs; in 1985–6 we again had a bird's-eye view of a pole, this time the south one. By 2007 we will be back to having an equatorial presentation.

Nobody has any real idea why Uranus is tipped in this way. The favoured theory is that at an early stage in its career it was struck by a massive body and literally knocked sideways. I admit to being profoundly sceptical; I fail to see how a largely liquid body over 30,000 miles across could be jolted in such a way, but I am also bound to admit that I cannot think of anything better – though other factors, to which I will return later, indicate that something very unusual happened in the outer Solar System several thousands of millions of years ago.

Even large telescopes will show virtually nothing on Uranus' pale disk. In 1992 I had the chance to observe it with the 60-inch reflector at Palomar, in California, and to me the planet was completely blank. I am unhappy about 'details' shown on drawings made by observers with much smaller instruments. Uranus is a very bland world, and clearly much less active than Jupiter or Saturn – or Neptune, for that matter.

Uranus has been suspected of short-term and long-term changes in brightness, possibly indicating changes in the outer clouds, although slight variations in the output of the Sun may also play a part. Amateur observations can be useful here, and can be carried out in the same way as for variable stars, but making accurate estimates is not easy, because Uranus shows a definite disk instead of appearing as a sharp point of light. I found this in May 1955, when Uranus was close to Jupiter in the sky. From 8 to 11 May the two planets were in the same telescopic field, and I decided to estimate the magnitude of Uranus by comparing it with the Galilean satellites of Jupiter, using various magnifications on my 12½-inch reflector. Alas, I failed. Uranus looked larger than any of the Galileans, but it was also much dimmer, area for area, and I found that I was quite unable to make any reliable comparisons. (It is hardly necessary to add that the two planets were not genuinely close together; they simply happened to lie in the same direction as

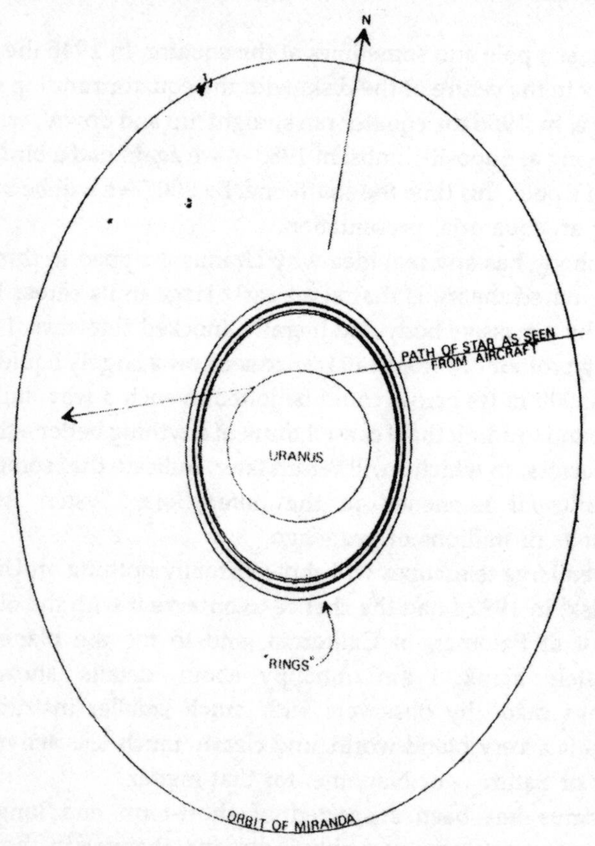

Discovery of the rings of Uranus: 10 March 1977. The star (SAO 158687) was occulted by Uranus, but 'winked' symmetrically both before and after actual occultation by the disk of the planet.

seen from Earth. Uranus is more than three times as far away from us as Jupiter.)

Another way in which the amateur can help is by observing occultations of stars by Uranus, the problem being that because Uranus moves so slowly occultations are rare. However, one occurred in 1977, and led to a most important discovery.

The date was 10 March; the star concerned was of the eighth magnitude. The occultation was observed from various stations, including the Kuiper Airborne Observatory, which is an aircraft specially modified to carry a large reflecting telescope. Both before and after the occultation the star 'winked' several times. The only possible explanation was that the star was being briefly hidden by dark rings associated with Uranus. Subsequently the rings were actually observed, using special infra-red techniques, and even before the Voyager 2 mission we had found out a good deal about them. The system is quite extensive, but there is no comparison with the glorious rings of Saturn, which are brilliant and icy; the Uranian rings are as black as coal-dust, and they are very narrow.

Five satellites – Miranda, Ariel, Umbriel, Titania and Oberon – were known in pre-Voyager days. All are smaller than our Moon, and even Titania, the largest, has a diameter of no more than 981 miles, so that it is comparable with Rhea and Iapetus in Saturn's family. It was generally expected that all the satellites would be icy and cratered. Whether or not Uranus had a magnetic field, and whether it sent out radio waves, had to wait for the Voyager results.

Voyager 2 had had a long flight since leaving the neighbourhood of Saturn in 1981. Moreover, it had not been trouble-free. The scan platform carrying the main camera had stuck at the very end of the Saturn encounter, and for a time it was feared that it would never be freed; apparently it had not been sufficiently lubricated. However, on arrival at Uranus all was well, and Voyager performed faultlessly throughout. This was a new sort of encounter, because the space-craft would approach its target pole-on, so to speak; it was rather like aiming for a bull's-eye in the centre of a dartboard, the bull, in this case, being Uranus' southern, sunlit pole.

The first major discovery came on 30 December 1985, almost a month before closest approach. Voyager located a new satellite,

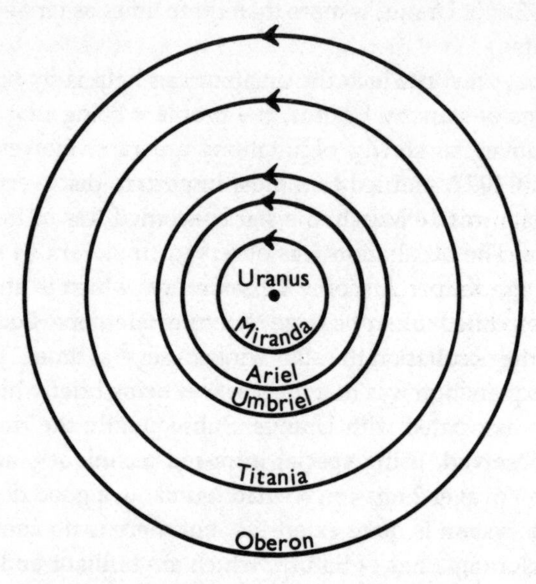

Orbits of the principal satellites of Uranus. All the newly-discovered satellites are within the orbit of Miranda.

closer to Uranus than Miranda, the innermost of the previously-known moons; following the Shakespearian tradition, it has been named Puck. Nine other small satellites followed, and one astronomer commented that it looked as if the Almighty had taken a shaker and scattered satellites in all directions. All were midgets; even Puck, the largest of them, is less than a hundred miles across. Voyager managed to obtain a picture of it, showing it to be darkish and crater-scarred.

The larger satellites were also icy, but were not alike. Oberon's craters tended to have darkish floors; Titania showed not only craters but also valleys and ice-cliffs; Umbriel was more subdued, and seemed to have an older surface, though there was one feature, close to the limit of Voyager's view, which was brighter and might possibly have been seen as a crater if it had been better placed (it has been named Wunda). Ariel has broad, branching valleys which look very much as though they have been cut by liquid, but all the satellites are too small to retain any trace of atmosphere, so that it is not easy to see how liquid water can ever have flowed on Ariel. Miranda was the showpiece of the system. It has an immensely varied surface; there are cratered plains, bright areas with cliffs and scarps, and 'coronæ', large trapezoidal-shaped regions which look rather like racetracks. Miranda itself is less than 300 miles across, and its strange surface is very much of a puzzle. It has been suggested that early in its career it was broken up by collision with a larger body, and re-formed later, but we do not really know.

The rings were well seen. There are ten of them altogether, plus a broad sheet of tenuous material stretching down from the innermost ring almost as far as the cloud-tops. The broadest ring is the outermost, known as the Epsilon ring, which has two 'shepherd' satellites, Cordelia and Ophelia, which are of course Voyager discoveries. The last picture, taken as Voyager was drawing away from Uranus, shows that there is a great deal of dust in the ring-system; the rings themselves seem to be made up of particles a few feet across, and they cannot be as much as one mile thick.

Few clouds were seen as Voyager 2 closed in. There were none of the disturbances seen on Jupiter or Saturn, and Uranus

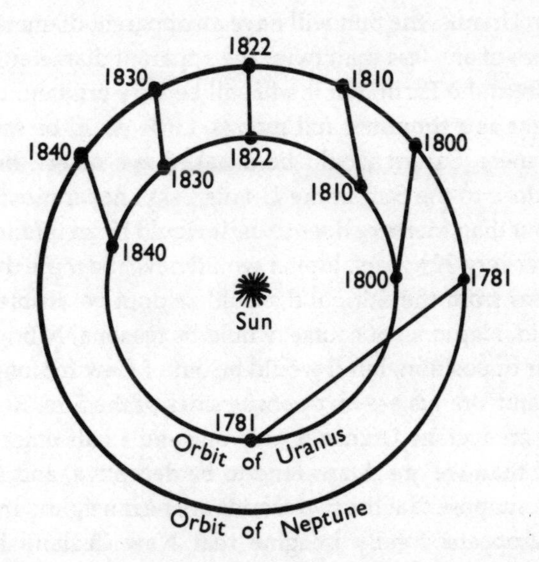

Pull of Neptune on Uranus. Before 1822 Neptune tended to accelerate Uranus; after 1822, to retard it.

appeared to be depressingly featureless. Eventually some obscure clouds were seen, and radio emissions were picked up, together with indications of a magnetic field. It was then found that the magnetic field is reversed relative to ours, so that what we call the north pole of rotation is magnetic south, but in any case the magnetic axis is inclined to the rotational axis by almost 60 degrees, and does not even pass through the centre of the globe.

This is certainly very odd, and was quite unexpected. It means, incidentally, that Uranian auroræ are much more conspicuous near the globe's equator than near the poles of rotation. The magnetosphere spreads out to 370,000 miles on the day side of the planet and 3,700,000 miles on the dark side, which means that all the members of the satellite family are immersed in it. Observations made at short wavelengths showed strong emissions on the day side, producing what is now termed an 'electroglow', unlike anything else we have so far found in the Solar System.

In many ways Uranus is the 'odd one out' among the giant planets. It alone lacks any marked internal heat-source; it alone has an extreme axial tilt; it shows very little surface activity – and there is virtually no difference in temperature between the equator and the poles.

From Uranus, the Sun will have an apparent diameter of only $1\frac{1}{2}$ minutes of arc, less than twice the apparent diameter of Jupiter as seen from the Earth, but it will still be very brilliant, casting as much light as a thousand full moons. Little could be seen of the other planets. Saturn would be a naked-eye object, but would remain close to the Sun in the Uranian sky, never moving much further out than Mercury does to us; it would be an inferior planet, best seen every $22\frac{3}{4}$ years. Jupiter would never be more than about 17 degrees from the Sun, and would seldom be visible without optical aid. Neptune, of course, would be reasonably bright when at or near opposition, but it would be out of view for long periods when it and Uranus are on opposite sides of the Sun. Remember, too, that on average Uranus is only one and a half times closer to Neptune than we are. Maps tend to be deceptive, and it is quite wrong to suppose that the two worlds are near neighbours – just as some Europeans fondly imagine that New Zealand is almost within hailing distance of Australia!

It is by no means difficult to identify Uranus. Between 1989 and 1995 it remains in Sagittarius, the Archer; it then moves into Capricornus, the Sea-goat, and will remain there until the end of the century. Binoculars will show that it is unlike a star, and a telescope will magnify it into its tiny bluish-green disk. It is an interesting and strange world, and it also has the distinction of being the first planet to be discovered by modern man.

CHAPTER FOURTEEN

Neptune

Far away in the depths of the Solar System, a thousand million miles beyond Uranus, lies the last of the giant planets – Neptune. Neptunian astronomers, if they existed, could know nothing about the Earth; but strangely enough, Earth astronomers knew about Neptune before they actually observed it.

The key was provided by Uranus, which had been recorded several times before Herschel identified it in 1781; Flamsteed's first record of it was made as early as 1690, so that observations extended back for almost a century, which is longer than one Uranian 'year'. It should, therefore have been possible to compute a really accurate orbit. Unfortunately, the old observations did not seem to fit in properly with those made after 1781. Something was wrong somewhere, and eventually a French mathematician, Alexis Bouvard, rejected the old observations altogether and worked out a new orbit, using only positions measured after Uranus had been recognized as a planet.

Even this would not do. Uranus refused to behave; it persistently strayed from its predicted path. Up to 1822 it seemed to move too quickly, while after 1822 it lagged. It became clear that there must be some unknown factor to be taken into account.

In 1834 the Rev. T. J. Hussey, Rector of Hayes in Kent, made a most interesting suggestion. Suppose that an unknown planet were pulling on Uranus? This might account for the irregularities in motion, and by 'working backwards', so to speak, it might be possible to track down the culprit.

Hussey went so far as to write to George (afterwards Sir George) Airy, who became Astronomer Royal at Greenwich in 1835. Airy was not encouraging, and replied that theory 'was not yet in such a state as to give the smallest hope of making out the nature of any external action' on Uranus. Hussey may have felt rebuffed; at any rate he took the matter no further, and the next step came from Eugène Bouvard, Alexis Bouvard's nephew, who in 1837 exchanged letters with Airy – who gave his opinion that even if an unseen body were responsible, 'it will be nearly impossible ever to find out its place'. Yet the problem of Uranus would not go away, and it was taken up again in 1841 by a young Cambridge undergraduate, John Couch Adams. He wrote in his diary:

> 'Formed a design, at the beginning of the week, of investigating as soon as possible after taking my degree, the irregularities in the motion of Uranus, which are as yet unaccounted for; in order to find whether they may be attributed to the action of an undiscovered planet beyond it; and if possible thence to determine the elements of its orbit, etc. approximately, which would probably lead to its discovery.'

He did pass his degree – brilliantly – in 1843, and from that time onward the movements of Uranus were very much in his thoughts. By October of the same year he had completed most of his research, and by the middle of 1845 he had derived an approximate position for the new planet. All that had now to be done was to look for it with a telescope.

Adams was not a practical observer, at least at that stage of his career,* and he cast around for help. He had already been in touch with James Challis, professor of astronomy at Cambridge, and now he again wrote to Airy. This was the start of a series of misfortunes which led to a most undignified dispute in after years. Airy, partly through a lack of confidence in the work of a young and unknown mathematician, took no action; Adams called to see him twice, but on the first occasion Airy was away, and on the second visit the butler told Adams that the Astronomer Royal was having dinner and could not be disturbed. (This may seem odd, because the time was 3.30 in the afternoon, but Airy had a rigid personal schedule which was never altered. Probably he did not

* When writing my book about Neptune I contacted Adams' descendants, and confirmed that at the time he did not even have access to a good telescope. See *The Planet Neptune*, published by Ellis Horwood (Chichester, 1988).

know that Adams had called.) Adams did not try again, but left a letter in which he gave the distance of the hypothetical planet as 38.4 astronomical units, which agreed with Bode's Law.

Airy replied in November, but unfortunately he asked a question which Adams regarded as trivial, and again nothing was done. It must be said that Airy, great astronomer though he undoubtedly was, was obsessed with 'order and method', and once he had made up his mind he was very reluctant to change his views. (It is said that on one occasion he spent a whole day in the cellars of Greenwich Observatory labelling empty boxes 'empty'.) Meanwhile, there had been developments on the other side of the Channel.

Urbain Jean Joseph Le Verrier, a young French mathematician, had taken up the problem of Uranus in very much the same way as Adams had done – naturally without any knowledge of Adams' work, because nothing had been published. Le Verrier approached matters differently, and published two memoirs, one in 1845 and the second in 1846. When Airy saw the second memoir he realized that Le Verrier's results were almost exactly the same as those of Adams, and the hunt for the new planet was on.

Airy, of course, was the director of Britain's leading observatory as well as being Astronomer Royal, and he might have been expected to start a personal search. This is precisely what he did not do. There was no suitable telescope at Greenwich, and in any case Airy was disinclined to upset the regular routine. He did contact Challis, and instructed him to begin searching with the powerful Northumberland refractor at the University. Challis obeyed, but with a strange lack of energy; neither did he have any good star-charts of the area, so that he had to work by purely visual means which were both time-consuming and laborious.

Le Verrier sent his results to the Paris Observatory. Nothing was done. Patience was never Le Verrier's strong point, and he sent his papers to Johann Galle of the Berlin Observatory, asking him to look in the indicated position. Galle was very ready to do so, and together with a young assistant, Heinrich d'Arrest, he set to work. He was fortunate in having an excellent telescope, and also an up-to-date star-map; moreover he had complete faith in Le Verrier's work. Sure enough, the planet was located on the very first night of observation. It showed a small but distinct disk, and it moved appreciably over a period of a few hours.

Johann Encke, director of the Berlin Observatory, lost no time in making the discovery known. On 28 September 1846 he wrote to Le Verrier: 'Allow me, Sir, to congratulate you most sincerely on the brilliant discovery with which you have enriched astronomy. Your name will forever be linked with the most outstanding conceivable proof of the validity of universal gravitation, and I believe that these few words sum up all that the ambition of a scientist can wish for it. It would be superfluous to add anything more.'

Challis was still plodding on at Cambridge, unaware that he was no longer alone in the hunt. When he finally learned of Le Verrier's triumph, he checked back on his own observations and found that he had recorded the planet twice during the first four nights of observing. Had he compared his notes, he could not have failed to make the discovery!

When the French learned that Adams had reached the same result as Le Verrier, and had completed his calculations much earlier, they were furious. It was felt that the English were trying to steal the glory of the discovery, and the whole affair very nearly blew up into an international incident. Fortunately neither Adams nor Le Verrier took any part in the dispute, and when they met face to face they struck up an immediate friendship, even though Adams could not speak French and Le Verrier was equally unfamiliar with English. After some discussion the new planet was named Neptune, after the Olympian sea-god.

As soon as Neptune was discovered, the orbit of Uranus was re-calculated, and this time the old observations fell almost perfectly into place. So far as Uranus was concerned, Neptune was at opposition in 1882; before that date, therefore, Neptune was tending to pull Uranus along, while after 1822 the effect was reversed. Had Uranus and Neptune been on opposite sides of the Sun during the early nineteenth century the perturbations on Uranus would have been almost inappreciable, and – apart from sheer chance – the discovery of Neptune would have been considerably delayed. The orbital period is 164.8 years, and, as we have seen, Neptune does not conform to Bode's Law.

There is one strange twist to the story. When Galileo made his observations of the four large satellites of Jupiter, in January 1610, he also sketched in the positions of adjacent stars, and there is no doubt that one of these 'stars' was Neptune. Galileo even noted that it had shifted in position, but he can hardly be blamed for not recognizing it as something new.

Neptune proved to be almost the same size as Uranus. It is in fact very slightly smaller, but appreciably denser and more massive. No Earth-based telescope can show much on its blue disk; I have looked at it through the Palomar 60-inch reflector without being able to see any detail, and I doubt whether even a keener-eyed observer would have been more successful, though modern photographs taken in light of short wavelength do show definite patches.

As soon as Neptune was found it was observed by the well-known English amateur William Lassell, whose telescope was one of the best in Europe. Lassell suspected a faint ring, but this later proved to be an optical effect; the real ring system was not discovered until the Voyager 2 mission of 1989. However, Lassell

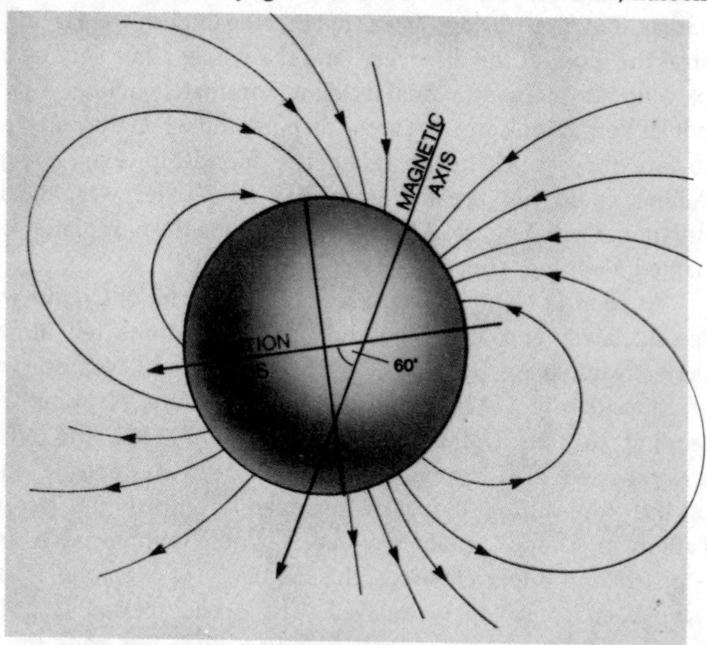

Inclination of Uranu's magnetic axis to its rotational axis.

did discover the large satellite Triton, which has an almost circular orbit but – uniquely among major planetary satellites – has retrograde motion. The second pre-Voyager satellite, Nereid, was not found until 1949, when it was discovered by G. P. Kuiper from the McDonald Observatory in Texas. Nereid is very small, and has a highly eccentric orbit more like that of a comet than a satellite; the distance from Neptune ranges between 836,000 miles and over 6,000,000 miles, in a period of 360 days.

Various other facts were established. Neptune does not share Uranus' remarkable tilt; the axial inclination is only five degrees greater than that of the Earth. The length of the rotation period was hard to establish because of the lack of visible detail, and only after the Voyager pass was it finally fixed at 16 hours 7 minutes.

Uranus and Neptune may be twins, but they are non-identical. Neptune, unlike Uranus, has a strong internal heat source. It was therefore expected to be a more active and dynamic world than Uranus – and so it proved.

At 0650 GMT on 25 August 1989, Voyager 2 passed over Neptune's darkened north pole at only a little over 3000 miles above the upper clouds – a closer encounter than for any of the

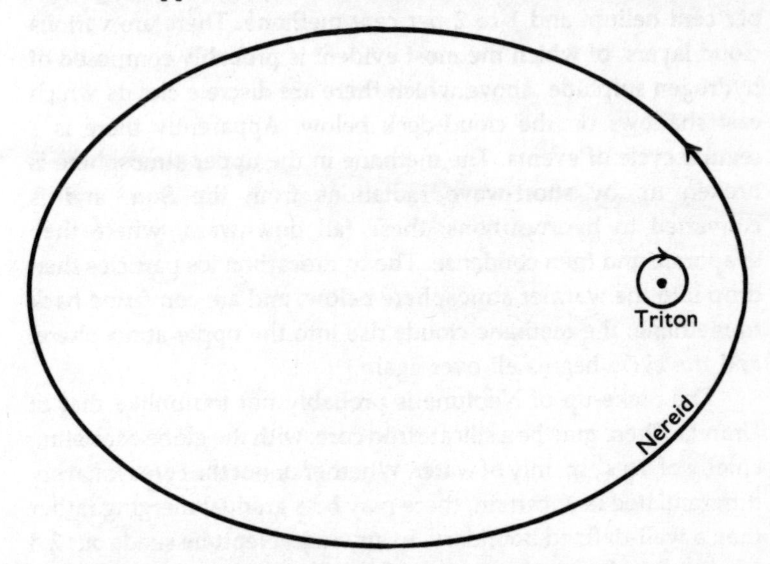

Orbits of Triton and Nereid. Triton has retrograde motion; Nereid direct. All the newly-discovered satellites are closer in than Triton, so that it is impossible to show them on this scale.

other giants. The spacecraft performed perfectly throughout the mission, even though it had been in flight for over twelve years and had covered over 4000 million miles. The pictures sent back were every whit as good as those obtained of Jupiter in 1979.

The main feature on the disk was a huge oval, now known as the Great Dark Spot; in size it is about the same, relative to Neptune, as the Great Red Spot is relative to Jupiter, and the latitude also is much the same. It drifts westward relative to the adjacent clouds, and it is rotating in an anti-clockwise direction; above it are wispy clouds made up of methane crystals, and known as methane cirrus. Further south there is a smaller, very variable feature which has a quicker rotation period, and has become known as the Scooter; still further south there is another dark spot, D2, which 'laps' the Great Dark Spot once in every five Earth days. Neptune is a decidedly windy planet, with velocities of up to 700 miles per hour and, as with the other giants, the rotation period is shortest at the equator and longest at the poles. The temperature is much the same as that of Uranus; the inner heat-source compensates for Neptune's much greater distance from the Sun.

The upper atmosphere is made up of 85 per cent hydrogen, 13 per cent helium and 1 to 2 per cent methane. There are various cloud layers, of which the most evident is probably composed of hydrogen sulphide, above which there are discrete clouds which cast shadows on the cloud-deck below. Apparently there is a regular cycle of events. The methane in the upper atmosphere is broken up by short-wave radiations from the Sun, and is converted to hydrocarbons; these fall downward, where they evaporate and then condense. The hydrocarbon ice particles then drop into the warmer atmosphere below, and are converted back to methane; the methane clouds rise into the upper atmosphere, and the cycle begins all over again.

The make-up of Neptune is probably not too unlike that of Uranus. There may be a silicate iron core, with the globe consisting chiefly of 'ices', mainly of water. Whether or not the core is sharply differentiated is uncertain; there may be a gradual merging rather than a well-defined boundary. In any case, Neptune sends out 2.8 times as much energy as it would do if it depended entirely upon what it receives from the Sun, and this explains why the temperature is no lower than that of Uranus.

Neptune was found to be a radio source, which was only to be expected. The real surprise was that the magnetic field proved to be almost as sharply tilted as that of Uranus; the angle between the rotational axis and the magnetic axis is 47 degrees, and, as with Uranus, the magnetic axis does not pass through the centre of the globe. It has been thought that the curious position of Uranus' magnetic axis was due to the great rotational tilt, but apparently this is not so, and the mystery remains.

Pre-Voyager observations of occultations of stars by Neptune had led to the suggestion that there might be incomplete rings – ring arcs, in fact. Voyager 2 found that there are five complete rings, much more obscure than even those of Uranus. They are not regular; the main ring has brighter sections in it. If all the ring material could be brought together, it would form a satellite no more than three miles across.

It had been expected that new small satellites would be found, and Voyager located six: Naiad, Thalassa, Despina, Galatea, Larissa and Proteus. Much the largest is Proteus, with a diameter of 258 miles; it is actually larger than Nereid, but cannot be seen from Earth because it is too close to Neptune. One Voyager image was obtained of it, and shows a rugged, crater-scarred surface. Galatea moves very close to one of the rings, and is probably a 'shepherd', though no other ring-shepherds were found despite a careful search. All the new satellites are closer in than either Triton or Nereid.

Just over five hours after passing over Neptune's north pole, Voyager 2 encountered its very last target – Triton, which proved to be an amazing world. It is smaller than had been believed, with a diameter of only 1681 miles, so that it is rather smaller than our Moon (some earlier estimates had indicated that it might be almost as large as Mars!). There had been suspicions that the surface might be cloud-hidden, but this is not so, and Triton's atmosphere is too thin to produce anything more noticeable than haze. The globe seems to be made up of more rock and less ice than the medium-sized and small satellites of Saturn and Uranus, and the surface is intensely cold; at a temperature of −400 degrees Fahrenheit (−236 degrees Centigrade) Triton is much the chilliest place ever encountered by a man-made probe.

Triton's surface seems to be covered with a general coating of ice, presumably water ice, overlaid by nitrogen and methane ices. Water ice has not been detected spectroscopically, but it must exist, because nitrogen and methane ices are not strong enough to maintain surface relief over long periods – they tend to flow. Not that there is much surface relief on Triton, and certainly there are no mountains, so that the difference in height between the loftiest and the lowest areas cannot be more than a few hundred feet.

The sunlit south pole is pink, due to nitrogen snow and ice. The colour is striking, and there were curious streaks here and there which at first defied explanation. Ordinary craters are scarce, but there are extensive flows, probably due to ammonia-water fluids which have frozen. At the edge of the pink polar cap there is a bluish area, caused by tiny crystals of methane ice; closer to the equator there is what has been termed 'cantaloupe' terrain because it has been said to look like the skin of a melon, with long fissures and subdued circular depressions. Elsewhere there are troughs, together with strange, mushroom-like features which are sometimes called guttæ, and whose origin is very uncertain. There are also frozen 'lakes' with flat, smooth interiors, probably due to water ice.

The region of the pink polar cap, Uhlanga Regio, contains the dark streaks. It seems that below the frozen surface there is a layer of liquid nitrogen. If for any reason this nitrogen starts to rise through the crust, there comes a point where the pressure has relaxed too much to keep it as a liquid, so that it literally explodes in a shower of nitrogen ice and vapour; the material is then blown along in the thin atmosphere. In this case, the streaks are nothing more or less than geysers, and Triton is an active world – the last thing that anybody had expected. Another explanation is that dust particles on the surface trap the sunlight and raise the temperature below them to above the boiling point of nitrogen, but either way the geyser idea seems to be valid. The material can rise as high as five miles, and can be blown downwind for as much as a hundred miles. The Tritonian atmosphere is a mixture of nitrogen and methane, with a pressure of only 1/70,000 of that of the Earth's air at sea level.

In all probability Triton was not originally a satellite of Neptune, but was once an independent body. When first captured by Neptune its orbit would have been elliptical, but over a period of around a thousand million years the path would be forced into a circular form; during this time the interior was churned around and heated (as Io's is today) and material flowed out on the surface, finally freezing there. Certainly Triton, with its pink snow and its nitrogen geysers, is unlike any other world in the Solar System.

What we would like to do, of course, is to take another look at Triton in the foreseeable future. Its seasons are immensely long and complicated, and there are bound to be marked changes in the distribution of the ice. Nitrogen ice can flow in the same way as a glacier, and it may even migrate from one pole to the other. Unfortunately no more probes to the outer Solar System are planned as yet, and we may have a long wait before we learn more. Moreover, the eccentrically-orbiting satellite Nereid was in the wrong position to be well shown by Voyager 2, so that we know very little about it.

From Neptune, the sunlight would be equal to almost 700 full moons – that is to say, eighty times the light of an ordinary candle seen from a distance of three feet. The maximum elongation of Venus from the Sun would be a mere 1½ degrees, Earth 2 degrees, Mars 3 degrees and Jupiter 10. Saturn would be a naked-eye object when best placed, though it must be remembered that Saturn is further away from Neptune than we are from Saturn. Even Uranus would be out of view for long periods, so that Neptunian astronomers would indeed have a poor view of the other planets.

With Neptune we come to the border of the main Solar System. Pluto remains; but at the moment Neptune ranks as the outermost known member of the planetary family.

CHAPTER FIFTEEN

Pluto

With the discovery of Neptune, the Solar System seemed to be complete once more. The wanderings of Uranus had been explained; the old observations by Flamsteed and others had fallen into place, and all the irregularities which had so puzzled Alexis Bouvard had been cleared up. Such was the general opinion for many years. And then – very slowly, very slightly – the outer giants started to wander again.

The differences between the observed and the theoretical positions were so tiny that they might easily have been put down to errors in measurement, but uneasy doubts remained. Could there be yet another planet, still further away from the Sun in the far reaches of the Solar System? One man who thought so was the American astronomer David Peck Todd, and in 1877 he began a systematic search with the 27-inch telescope at the United States Naval Observatory. He hoped to find an object which would show a small disk. He failed, and so did various other hunters, but as time went by it began to look more and more likely that there really was a ninth planet waiting to be tracked down.

At this point Percival Lowell took up the problem. As we know, Lowell built his observatory at Flagstaff mainly to study Mars, and he is best remembered today because of his belief in canal-building Martians, but this is a pity, because his contributions to astronomy were very great. He was an expert mathematician, and at the turn of the century he began to make calculations, concentrating mainly upon the movements of Uranus simply because they were better known than those of Neptune. Remember that Neptune was discovered only in 1846, and had not completed a single journey round the Sun since it had been identified (in fact, it has not done so even yet). Lowell decided that

'Planet X' had a period of 282 years and a mass seven times that of the Earth; the orbit, he believed, was decidedly eccentric, and the next perihelion would fall in 1991. He worked out a position, and then started looking.

He continued the search between 1905 and 1907, using the great Lowell refractor. In one way his task was easier than that of Adams and Le Verrier so long before, because he could use photography. On the other hand Planet X was bound to be much fainter than Neptune, and the calculated position was very uncertain because it was based upon such tiny irregularities in the motion of Uranus. Under the circumstances it was hardly surprising that the planet refused to show itself, and a second search from Flagstaff, carried out in 1914 by C. O. Lampland, was equally fruitless. Lowell died suddenly in 1916, and for a while the problem of Planet X was shelved.

It surfaced again in 1919, when Milton Humason, at the Mount Wilson Observatory, undertook a photographic search on the basis of calculations by another American astronomer, W. H. Pickering. Pickering's method was rather different from Lowell's, and he decided to concentrate upon Neptune rather than Uranus. He also had another clue, provided by those flimsy and erratic wanderers, the comets.

Although some comets are of vast size, their masses are negligible even compared with a small satellite such as Phœbe, and their orbits can be violently perturbed by the gravitational pulls of the planets. Many of the periodical comets have their aphelia at about the same distance as the orbit of Jupiter; this is no mere coincidence, and astronomers always refer to Jupiter's comet family. Pickering pointed out that there were sixteen known comets with their aphelion distances at about 7,000,000,000 miles from the Sun, and this made him even more certain that there must be a planet there. His result was very similar to Lowell's, but Humason was no more successful than the Flagstaff team had been, and once more the problem was put 'on hold'.

The next step was taken in 1929. Lowell's assistant, V. M. Slipher, had become director of the observatory at Flagstaff, and he was determined not to let Planet X elude him. He obtained a 13-inch refracting telescope specially for the search, and enlisted the aid of a young amateur, Clyde Tombaugh, who had submitted some very impressive drawings of Mars and the other planets. Tombaugh arrived at the observatory and began work.

His method was essentially the same as Lowell's. The same region of the sky was photographed twice, with an interval of several days between the exposures; the stars would stay in the same relative positions, but a planet would move. The two plates were then compared in an ingenious instrument called a blink-microscope, and a moving object would be seen to 'jump'.

Tombaugh met with success earlier than he had dared to hope. Plates exposed on 23 and 29 January 1930 showed a dim speck which moved at just the right speed for just the right distance. As Tombaugh checked with the blink-microscope, he recorded that 'Suddenly I spied a 15th-magnitude image moving out and disappearing in the rapidly alternating views. Then I spied another image doing exactly the same thing about three millimetres to the right of it. "That's it," I exclaimed to myself.'

It was indeed the new planet. Over the next nights the Lowell astronomers checked and re-checked to make sure that there had been no mistake; finally, on 13 March – the seventy-fifth anniversary of Lowell's birth and the 149th of Herschel's discovery of Uranus – Slipher sent out a telegram to all the major observatories: 'Systematic search begun years ago supplementing Lowell's investigation for Trans-Neptunian planet has revealed object which since seven weeks has in rate of motion and path consistently confirmed to Trans-Neptunian body at approximate distance he assigned.' The difference between the actual position and Lowell's forecast was less than 6 degrees. Lowell had been right, too, in assuming that the planet would have a very eccentric and inclined orbit. The revolution period proved to be 248 years, and the next perihelion passage would fall in 1989.

The first thing to be decided was a name, and various suggestions were made. One was Minerva, after the goddess of wisdom, and this might well have been chosen had it not been proposed by T. J. J. See, an American astronomer who was not exactly popular with his colleagues, one of whom (A. E. Douglass) wrote that 'Personally, I have never had such aversion to a man or beast or reptile or anything disgusting as I have had to him. The moment he leaves town will be one of vast and intense relief, and I never want to see him again. If he comes back, I will have him

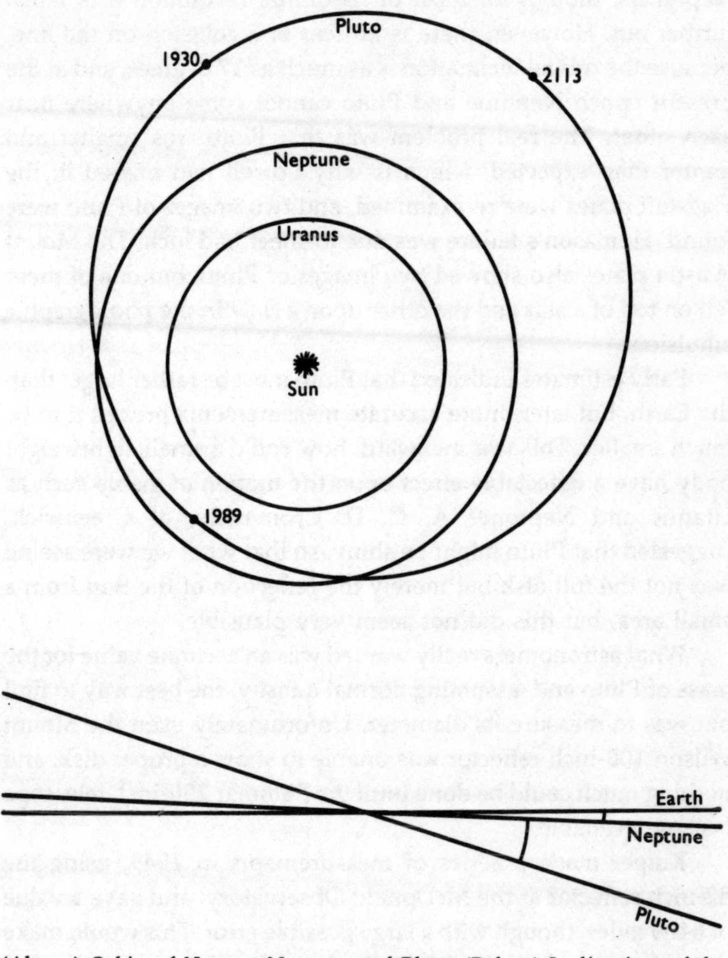

(Above) *Orbits of Uranus, Neptune and Pluto.* (Below) *Inclinations of the orbits of Neptune and Pluto.*

kicked out.' The name finally selected was Pluto, after the god of the underworld, proposed by an English schoolgirl, Venetia Burney. It was surely appropriate; Pluto's planet is a frigid, dimly-lit place.

Problems arose almost at once. Pluto's orbit was very strange; at perihelion the planet can come closer in to the Sun than Neptune, and between 1979 and 1999 its distance is less than

Neptune's, though for most of its orbital revolution it is much further out. However, there is no fear of a collision on the line, because the orbital inclination is as much as 17 degrees, and at the present epoch Neptune and Pluto cannot come anywhere near each other. The real problem was that Pluto was smaller and fainter than expected, which is why Lowell had missed it; the Flagstaff plates were re-examined, and two images of Pluto were found. Humason's failure was due to sheer bad luck. The Mount Wilson plates also showed two images of Pluto, but one of them fell on top of a star and the other upon a flaw in the photographic emulsion.

Early estimates indicated that Pluto must be rather larger than the Earth, but later, more accurate measurements proved it to be much smaller. This was awkward; how could a small, lightweight body have a detectable effect upon the motion of giants such as Uranus and Neptune? A. C. D. Crommelin, at Greenwich, suggested that Pluto might be shiny, so that what we were seeing was not the full disk but merely the reflection of the Sun from a small area, but this did not seem very plausible.

What astronomers really wanted was an accurate value for the mass of Pluto and, assuming normal density, the best way to find out was to measure its diameter. Unfortunately even the Mount Wilson 100-inch reflector was unable to show a proper disk, and nothing much could be done until the Palomar 200-inch telescope became available.

Kuiper made a series of measurements in 1949, using the 82-inch reflector at the McDonald Observatory, and gave a value of 6400 miles, though with a large possible error. This would make the mass about $\frac{8}{10}$ of that of the Earth, so that the perturbations of Uranus and Neptune might just be accounted for, allowing for observational errors and a few coincidences.

In March 1950, twenty years after Tombaugh's discovery, Kuiper and Humason, using the Palomar telescope, obtained new measurements which gave Pluto's diameter at a mere 3600 miles, less than that of Mars. Obviously this threw the whole question wide open. If Pluto were really smaller than Mars, and yet was massive enough to cause the perturbations in the movements of Uranus which led to its discovery, the density would have to be about twelve times that of the Earth, or sixty times that of water. In this case Pluto would be made up entirely of very heavy materials, and with a tremendous surface gravity, so that a man weighing 13 stone on Earth would weigh 60 stone on Pluto. This seemed wildly unlikely, and astronomers were frankly baffled.

One thing was certain. If Pluto were really small and made up of normal materials, it could not be massive enough to pull Uranus or Neptune perceptibly out of position. In this case Pluto was not the planet for which everyone had been searching. Either Lowell's correct prediction had been a sheer fluke, or else the real Planet X awaited discovery.

There were still doubts about Pluto's real size; it is surprisingly difficult to measure the apparent diameter of a very small disk which does not even have sharply-defined edges. An alternative method involved occultations. There are times when Pluto passes in front of a star, and the length of time for which the star is hidden tells us how big Pluto's disk must be. The method is excellent in theory, but it is limited because tiny, slow-moving Pluto is seldom co-operative enough to produce a suitable occultation.

An extra hazard was that although Pluto's path among the stars was reasonably well known, the slightest error would make a great difference in predicting occultations. Therefore, astronomers at the US Naval Observatory at Flagstaff began a long series of photographs of Pluto, taken with the large telescope at the Observatory. The disk was too small for the diameter to be determined with real accuracy, but the new measurements would, it was hoped, make it possible to predict future occultations suited for observation. But it was during these studies of the photographs that a completely unexpected discovery was made. Pluto is not a solitary traveller in space!

When the photographic images were examined, it seemed at first that Pluto might be either double, dumb-bell-shaped or simply irregular. Finally it was proved beyond a shadow of doubt that the effect was due to a second body, separate from Pluto and about half Pluto's size. The final proof came later from a magnificent photograph taken by the Hubble Space Telescope. The secondary body was named Charon, after the gloomy ferryman who rowed departed souls across the river Styx into Pluto's domain.

The distance between the two bodies, centre to centre, is 12,200 miles. The diameter of Pluto was at last determined very precisely; it is 1444 miles, while Charon measures 753 miles. The orbital period of Charon is 6 days 9 hours. Earlier, slight fluctuations in the magnitude of Pluto had given a rotation period of just this value – 6 days 9 hours – so that the two are 'locked'. From one hemisphere of Pluto, Charon would hang motionless in the sky; from the opposite hemisphere it could never be seen at all. To complete the strangeness of the situation, the rotational axis of Pluto is tilted by 122 degrees, so that in this respect it more nearly resembles Uranus than any of the other planets.

Nature can often be perverse, but in the mid and late 1980s she was kind. The orbits of Pluto and Charon were tipped so that for several years there were mutual occultations, with Pluto passing in front of Charon and hiding it, and Charon passing in front of Pluto and cutting out some of its light (much as the Moon does with the Sun during an annular eclipse). This gave observers an opportunity which will not recur for over a century. During occultations of Charon, Pluto's spectrum could be studied on its own; when Charon was in transit the two spectra could be seen together, but since that of Pluto was already known, it could be allowed for. Moreover, the ways in which the two moved made it possible to work out their masses very accurately. Pluto's mass turns out to be only about 18 per cent of that of the Moon.

All this means that we have a vastly better knowledge of the Pluto–Charon pair now than we did a few years ago. In 1980, exactly half a century after Pluto's discovery, a conference was held at Las Cruces, New Mexico, at which Clyde Tombaugh was the guest of honour; at that time our total information could be summed up by papers delivered in a single day, and we were not even quite sure that Charon was a separate body.

The two worlds are not alike. Pluto is much the more substantial of the pair, and is more than twice as dense as water, so that it contains less ice and more rock than the icy satellites of Saturn and Uranus. Pluto's surface is coated with methane ice, plus some nitrogen ice, while the coating of Charon seems to be ordinary water ice. Pluto has an extensive though very tenuous atmosphere, probably composed chiefly of nitrogen together with some carbon monoxide, while Charon has none – at least, none that we have so far been able to detect.

Though the atmosphere is extensive at the moment, despite its extreme thinness, it may not be present all through a Plutonian 'year'. At present the distance from the Sun is increasing, and the temperature is falling. During the coming century Pluto will become so cold that the atmosphere will freeze out on to the surface, and this will continue until the planet swings inward once more after reaching aphelion in the year 2113. Until we have a more detailed knowledge of the composition of the atmosphere we cannot be sure just when it will freeze out, but there seems every chance that it will happen. There could be an analogy here with Chiron,* which next reaches perihelion in 1995 and whose 'atmosphere' started to become evident only in 1988.

We even have a few hints of surface features. Apparently Pluto has a bright polar cap and a darker band across its equator, as we can tell from the changes in brightness when Charon passes over Pluto in transit.

There is no longer the slightest chance that Pluto is Lowell's Planet X, and it is also true to say that Pluto does not seem to be worthy of planetary rank – so what exactly is it?

An early suggestion, made by R. A. Lyttleton, was that it might be an ex-satellite of Neptune which broke away in some manner and moved off independently. This sounded reasonable enough, but the presence of Charon seems to rule it out, because if the two had originally been in orbit round Neptune they could not have remained locked in the way they are. On the other hand we have to consider Triton itself, which is larger and more massive than Pluto, and has a retrograde orbit. It may well be that Triton and Pluto are of similar type, and that Triton has been captured by Neptune while Pluto has remained free. Note, too, that both have tenuous nitrogen atmospheres.

Another idea is that Pluto and Charon may represent a case of a double asteroid. Even in the main asteroid belt there have been some indications of 'pairs', though as yet we have no proof; the trouble here is that Pluto, small though it may be by planetary standards, is still much too large to be an asteroid. Even Charon has a diameter greater than that of Ceres, much the senior member of the asteroid swarm.

* Do not confuse Chiron, the asteroidal body moving between the orbits of Saturn and Uranus, with Charon, the companion of Pluto.

It may be more plausible to class Pluto as a planetesimal, one of the fragments from which the planets built up in the original solar nebula. In this case Triton, Charon, Chiron, and irregular asteroids such as Pholus, 1992 QB1 and 1993 FW could also be planetesimals.

Pluto is within the range of good amateur telescopes, but there are only two possible avenues of useful research. One is to make magnitude estimates, using the same techniques as those of variable star observers; any variations will probably be too slight to be seen visually, so that a photometer will be needed. The other is by observing occultations. Here the amateur can be really useful, because he is more mobile than the average professional and can take equipment to suitable points on the Earth's surface – but, of course, the opportunities are so rare that there may be only one chance in a lifetime.

Pluto remains an enigma. It does not seem to be a planet; it is not a normal asteroid; we cannot be sure that it is a planetesimal. During the twenty-first century we may be able to study it from close range, and plans for a probe to it have already been drawn up, but when it will be launched we do not know. Pluto would certainly be worth visiting, even though it must be the loneliest and most desolate world that we can imagine.

CHAPTER SIXTEEN

Beyond the Planets

It is wrong to say that the Solar System ends at the orbits of Neptune and Pluto. Quite apart from the possible Planet X and maverick objects such as Pholus, there are the comets. The Kuiper Cloud – if it exists – is certainly further out than Neptune, and the distance of the Oort Cloud is more likely to be in the region of a light-year. Remember that the nearest stars, those of the Alpha Centauri group, are over four light-years away.

Clyde Tombaugh went on searching for a decade after he had found Pluto, and covered 70 per cent of the sky, reaching down to magnitude 17. Knowing him as I do, I refuse to believe that he could have missed 'another Pluto' had it been there, and this means that any new planet must be well below the seventeenth magnitude. But there could be several large planets at great distances – why not? – and this brings me on to the Oort Cloud, because I believe it may be relevant.

To recapitulate: The Oort Cloud is believed to be a reservoir of cometary bodies. If one of these bodies is perturbed, it will start to fall inward towards the Sun. One of several things may then happen to it. It may simply swing round the Sun and return to the Oort Cloud, not to be back for thousands or millions of years. (Delavan's Comet of 1914 has an estimated period of 24,000,000 years – give or take a few million years either way!) It may hit the Sun and be destroyed, as with Comet Howard-Kooman-Michels of 1979. It may be thrown out of the Solar System altogether, as with the 'spiked' comet Arend-Roland of 1957. Or it may be captured by the pull of a planet, usually Jupiter, and forced into a short-period orbit. Halley's Comet is the classic example; it returns to perihelion every 76 years, and each time it loses about 300,000,000 tons of material, so that by cosmical standards it

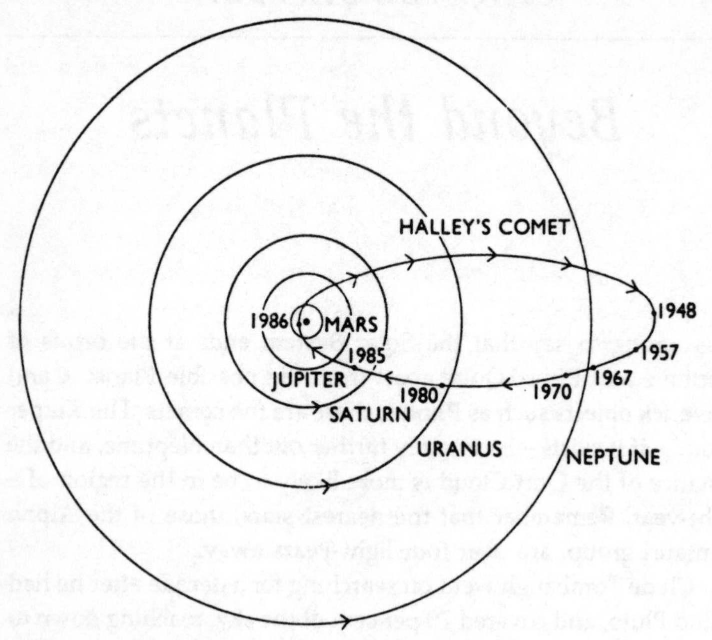

Orbit of Halley's Comet. It came to perihelion in 1986, and will next do so in 2061.

cannot be very old. There is certainly only a limited amount of mass to spare – it would take 60,000 million Halley's Comets to equal the mass of the Earth – and had it been in its present orbit for a very long time it would have lost all its volatiles. Therefore, it is a recent newcomer from the Oort Cloud.

There are many short-period comets, and several of them have 'gone missing' in modern times. Biela's Comet used to have a period of 6¾ years, but in 1846 it broke in half and has never been seen since 1852, though for some years we saw its débris in the form of a meteor shower each November (even this seems to be defunct now), while Westphal's Comet, with a period of 62 years, faded out during the return of 1913 and did not come back in 1975. Brorsen's Comet (period 5½ years) is another casualty; it was seen at five returns, but has been lost since 1879.

If short-period comets disappear so quickly, there must be continual replenishment of them from the Oort Cloud. But can passing stars affect the Cloud often enough? They are a long way away; the distance of Alpha Centauri is of the order of 25 million million miles. Close encounters between the Solar System and another star must be vanishingly rare. This is the problem.

Some astronomers believe that comets can be picked up from interstellar space, but in this case they would be expected to approach the Sun more quickly than they actually do, and it is more generally thought that they are bona-fide members of the Solar System, representing the material left over when the main planets were formed. My own suggestion – which may or may not be original; I have not seen it anywhere else, but I may have missed it – is that the perturbing influence on the comets in the Oort Cloud is not a wandering star, but a very remote and probably quite massive planet moving round the Sun either in a near-circular orbit, somewhere in the region of the Cloud, or else in a very eccentric or inclined one, in which case it would cross the Cloud only occasionally. Of course, the same could be said of the possible Kuiper Cloud.

With this aside, let us return to the possibility of finding the real Planet X.

Opinions differ, but my own view is that the accuracy of Lowell's prediction is really too great to be explained away as mere coincidence, in which case Planet X is really there. It is bound to be very dim, however, and the chances of our finding it unless we have at least a vague idea of its position are very slim.

Pickering's idea of a Planet X comet family was revived by K. Schütte in 1950; he believed that the planet lay at around 7000 million miles from the Sun. A photographic search was undertaken in the position given by Schütte, but with no result. Equally negative results were obtained in 1972 and 1975 by the Russian astronomer H. Chebotarev.

There was a flurry of excitement in 1972, when J. A. Brady, of the University of California, claimed that Halley's Comet was being perturbed by Planet X, and that the magnitude of the planet might be between 13 and 14, not very different from Pluto's. According to Brady, the planet was three times as massive as Saturn, and moved round the Sun in a retrograde orbit in a period of 464 years. The position he gave fell in the constellation of Cassiopeia, in the far north of the sky.

Observers such as myself flew to their telescopes, overhauled their cameras, and began a concentrated hunt. Again the results were negative, and then, rather to the general disappointment, new investigations showed that Brady's calculations were unsound. Certainly a large planet moving in a wrong-way direction would present theorists with serious problems!

Another line of investigation involves the four space-probes which are now on their way out of the Solar System, Pioneers 10 and 11 and Voyagers 1 and 2. We know exactly where they are, and we are monitoring their movements very precisely. If Planet X were within range, its effects could be detected. So far nothing of the kind has been found, indicating either that the planet does not exist, or else that it is out of range.

Yet when all is said and done, we come back to the possible effects of Planet X upon the two outer giants, Uranus and Neptune (we can forget about Pluto, because even now its orbit is not known with pinpoint accuracy). The most elaborate calculations have been made in America, by Robert Harrington of the United States Naval Observatory and James Anderson of the Jet Propulsion Laboratory in California. Harrington's work, based on the perturbations of Uranus, leads to a planet from two to five times the mass of the Earth, with a period of 600 years and a present distance of 6000 million miles from the Sun; the position given is in the region of Scorpius and Sagittarius – making a search even more irksome, because this is an area of the Milky Way where faint stars are very numerous. Anderson has taken both Uranus and Neptune into account, and concludes that there were genuine perturbations between 1810 and 1910, but not since. If Planet X is responsible, there can be only one explanation. A planet cannot 'softly and silently vanish away', like the hunter of the Snark, and so it must travel in a very eccentric orbit. Anderson gives it an inclination of almost 90 degrees, almost at right angles to the Earth's orbit, and a period of between 700 and 1000 years, with a mass five times that of the Earth.

It is all very nebulous, and we have to admit that the discovery of Planet X depends very largely upon luck – particularly if it is moving in an Anderson-type path, because in that case several centuries may elapse before it is again close enough to make its presence felt. Yet there is unquestionably something about the movements of the outer planets which remains unexplained, and to my mind at least Planet X is the most reasonable answer.

Various less plausible ideas have been proposed from time to time. One involves a massive planet which comes back periodically and sends comets hurtling inward from the Oort Cloud, so that there is a period of several hundreds of years when the Earth is under bombardment (this, needless to say, accounts for the extinction of the dinosaurs). A dead star companion of the Sun has also been proposed, and even given a name: Nemesis. Its distance is given as at least two light-years. But there is no proof, and, to be candid, theories of this kind are so highly speculative that to pursue them further at the present moment is rather pointless.

Planet X ought to be found one day. What it will be like we do not know. It may be a giant, and at least it ought to be massive enough to explain the tiny perturbations of Uranus and Neptune, but for the present time this is as far as we can go.

CHAPTER SEVENTEEN

Life on the Planets

Everyone is interested in the possibility of life beyond the Earth. In science fiction, aliens are usually represented as grotesque and decidedly unfriendly – a legacy, perhaps, from the Martians in H. G. Wells's classic *War of the Worlds* – but I have never understood why this should necessarily be so. In any case, what forms could life take?

We are very vague about the origin of life, but at least we know how it is made up, and all living material known to us is based upon the properties of one atom: that of carbon. Carbon atoms have a remarkable ability to build up complex molecules, both with other carbon atoms and with different elements, and it is these highly complicated molecules which are essential for life. The only other atom which has something of the same power is silicon, which however is not nearly so efficient; we have no evidence that silicon-based life can exist.

It follows that life, wherever it may be found, must be based on carbon, unless unknown elements exist. This latter idea flies in the face of all the evidence. There are ninety-two naturally-occurring elements, and these form a complete sequence; there is no room for an extra one, and the heavier elements which have been manufactured in our laboratories are always unstable, so that for the moment we can forget about them. Moreover, spectroscopic analysis of remote stars and star-systems tells us that the elements there are just the same as the elements here. The light from the Pole Star or an immensely distant galaxy is due to substances familiar to us.

This being so, I feel that we are entitled to discount what are known popularly as BEMs or Bug-Eyed Monsters, made of gold, living on airless worlds and nibbling rocks for breakfast. They may be fun (in books and films, at least!), but they are not science.

Of course, it is always possible to claim that there is a major loophole which would open the way for BEMs, but in this case all of our modern science would be wrong, and this seems excessively unlikely. Given a set of facts, all that one can do is to take them and then interpret them as reasonably as possible. Let me give you an example. From my home on the end of Selsey Bill I believe that the Sun will rise in the east tomorrow morning. Of course, it may rise in the west for a change; but I doubt it, because all the evidence is to the contrary. For the same sort of reason, I do not believe in totally alien life-forms; once we start discussing them we are reduced to speculation which is completely pointless. Therefore I propose to consider only life of the sort we can understand.

Life on Earth can assume many forms. There is not much superficial resemblance between an amœba, a daffodil, a jellyfish and a man, but the essential common factor – dependence upon carbon – is there. Life, then, depends upon a suitable environment. We need a reasonably equable temperature, a suitable atmosphere, and a supply of water. If any of these essentials are lacking, then there will be no life.

It follows that most of the worlds in the Solar System can be ruled out of court straight away. Mercury, the Moon, and all planetary satellites except Titan either lack atmospheres or have atmospheres which are too thin to be of any use at all; Titan is far too cold, and in any case its atmosphere contains too much methane and virtually no free oxygen. Jupiter and the other giants have no visible solid surfaces, and in every way are hopelessly unsuitable – I cannot credit living creatures swimming around beneath the Jovian clouds. Venus is equally unsuited to any life; a crushing carbon dioxide atmosphere, a fiercely hot environment and clouds of acid do not make it an inviting place. So we come back, as always, to Mars.

Less than a hundred years ago a wealthy Paris widow, Mme Guzman, offered a large sum of money as a prize for the first man who established communication with beings on another world – Mars being excluded, because it was too easy! Then we had Lowell's canals, which were taken very seriously indeed. And in an early, pre-Mariner edition of this book I confess to having written that 'it seems overwhelmingly probable that living organisms exist on Mars, and are responsible for the famous dark areas'. All this has changed now. Mars does not meet our preliminary needs; the atmosphere is so thin that it corresponds to what we usually call a reasonable laboratory vacuum, and the cold, during nights at least, is intolerable. Moreover, the Viking landers found no sign of biological activity. True, there were only two landers, and there is certainly something odd about Martian chemistry, but the evidence as it stands at the moment is strongly against the existence of any life on the Red Planet. I suppose we

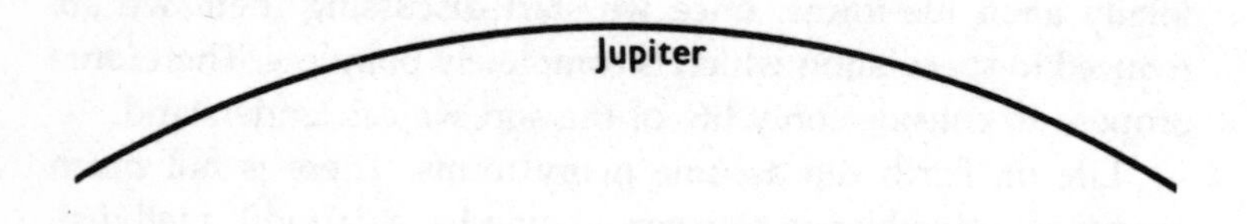

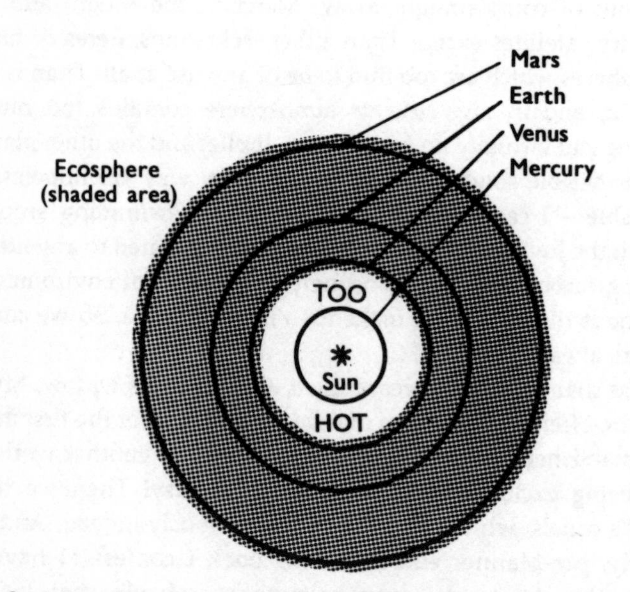

The Sun's Ecosphere. Venus is at the inner edge; Mars at the outer edge; only the Earth lies within it.

cannot finally rule out very primitive organisms, possibly underground, but we can certainly rule out anything as advanced as a beetle.

The crux of the matter is that the Earth, and only the Earth, moves in the region which is known as the solar ecosphere – that is to say the region in which the temperature is liable to be neither too hot or too cold, other things being equal. Venus is within the inner edge of the ecosphere, Mars at the outer boundary. Whether life has ever existed on either planet remains to be seen, but I am ready to go on record as saying that I doubt it. The whole evolution of Venus has been different from ours, because it is over 20 million miles closer to the Sun, and Mars presumably lost any atmosphere fairly quickly, giving life no time to gain a proper foothold – if any at all.

In the Solar System we are alone. There is no intelligent life except (possibly) on the Earth. Yet this is not the same as claiming that we are unique, which would be both parochial and illogical.

Our Galaxy contains 100,000 million stars, many of which are very like the Sun, and we have good reasons for supposing that many of these are attended by planetary systems. If we find a planet like the Earth moving round a star like the Sun, it seems reasonable to expect that life will have appeared there. I am quite prepared to believe that on, say, Delta Pavonis F there may be a highly intelligent astronomer with two heads and three arms. He may look like a bug-eyed monster, but he will not really be one, because he (or she, or it) will be carbon-based, and will need the same kind of environment.

Unfortunately, planets of other stars are so remote that they are out of contact with us. Sending a rocket even to Alpha Centauri is absolutely out of the question at present, and will probably remain so, since nothing can travel faster than light (186,000 miles per second), and any journey at speeds we can envisage would take an impossibly long time even if it were practicable in other ways. The only chance of establishing contact is by radio, since radio waves, which are electromagnetic vibrations, travel at the same speed as light.

In 1960 a team of scientists at Green Bank in West Virginia, headed by Frank Drake, carried out an experiment which would have seemed fantastic only a couple of decades earlier. We know that there are vast clouds of rarefied hydrogen spread between the stars of the Galaxy, and that these hydrogen clouds emit radiation at a wavelength of 21.1 centimetres. This 21.1-centimetre radiation can be collected by radio telescopes, such as that at Green Bank, and can tell us a great deal about the structure of the Galaxy. If there are other advanced beings anywhere, they will presumably know about the radio emission, and will be keeping a watch at this particular wavelength. The idea behind the Green Bank experiment was to try to detect signals at 21 centimetres which were sufficiently rhythmical to be classed as non-natural.

Drake concentrated upon the two nearest stars which are at all like the Sun, Tau Ceti and Epsilon Eridani. Alas, nothing was found. The experiment, known officially as Project Ozma but more commonly as Project Little Green Men, was discontinued after a time, but the radio telescope used for it is still in operation at Green Bank, and it will always remain of special historical interest.

Other attempts have been made since then, and one, started in 1992, is still going on; a concentrated effort is being made to 'listen out' in the hope of picking up a signal which cannot be dismissed as natural. The chances may be small, but they are not nil, and it is a measure of our changed attitude that the experiment is regarded as worth trying at all. It is significant that at the last General Assembly of the International Astronomical Union, held in Argentina in 1991, a whole day was devoted to the procedure to be followed by anyone detecting ETI (Extra-Terrestrial Intelligence).

Direct travel to other planetary systems is out of the question by our present techniques simply because nothing can move as fast as light,* and it is difficult to have much faith in space-arks, time warps, space warps, thought travel or teleportation; enter this realm, and we are cheek and jowl with Dr Who, Lord Darth Vader and the intrepid crew of the starship *Enterprise*. But though these methods may be fictional at the moment, we must bear in mind that television would have seemed equally absurd to anyone living a couple of hundred years ago or even less. It is fair to say that if we are to achieve interstellar travel, we must wait for some fundamental breakthrough which we cannot even discuss at the

* Anyone who doubts this must, I fear, delve into the abstruse mathematics of Einsteinian relativity!

moment. It may come this year, next year, in a hundred years, a million years – or never.

Of course, there is always the chance that beings from other worlds will come here, but the whole topic has been bedevilled by the flying saucer craze, which attracts cranks in the same way that a jamjar attracts wasps. Nobody will claim that such a visitation is impossible, because there must surely be many races in the Galaxy who are far more advanced technologically than ourselves, but there is no evidence that it has ever happened. If it does, then we need not be alarmed. Any race capable of making such a journey will long since have left war and conquest far behind it, and will come in peace.

Finally, what of the future?

The Moon has been reached; Mars must be next on the list, and a Martian Base in the foreseeable future is a strong possibility. Further than that we cannot predict as yet. Changing the atmosphere of Venus and turning it into a sort of second Earth calls for techniques hopelessly beyond those of our time, and the remaining planets and satellites do not look promising; but no doubt there will be many more unmanned missions, and by AD 2093 we should have a really good knowledge of every member of the Sun's family.

The Sun is a steady, well-behaved star. It has been shining much as it does now for several thousands of millions of years, and it is no more than middle-aged. Eventually, however, it will change. When it uses up its store of available hydrogen 'fuel' it will expand into a red giant star, and this must mean the end of the Earth. The inner planets will be scorched and probably destroyed; the outer planets will survive, but there is no chance that they will become habitable. In any case, the Sun's blaze of glory will not last for long on the cosmical scale, and will be followed by collapse into a very small, feeble, super-dense white dwarf star, bankrupt of all nuclear energy and unable to shed more than a dim radiance upon its remaining planets. The end of the Solar System must be a cold, dead sun attended by the ghosts of its surviving family.

This may sound depressing, but it lies so far in the future that we cannot claim to be at all confident about the sequence of events.

If men still live on the Earth at the time when the Sun starts to change, they may have learned enough to save themselves. But this is pure speculation, and at least we are confident that the Earth will remain habitable for an immense period in the future – always provided that we do not damage it irreparably.

Meantime, our task is to find out as much as we can about our neighbour worlds. The planets are our companions in space; they are no longer inaccessible, and they offer us the greatest challenge ever made to the ingenuity of mankind.

APPENDIX I

Observing the Planets

Before the Space Age, much of our knowledge of the surface features of the Moon and planets was based on amateur work. This is not true today, of course. There is still the same enjoyment in observing the planets and making drawings of them; who can tire of gazing at the rings of Saturn or the belts and moons of Jupiter? But the amateur who wants to carry out useful research has to be more specialized and better-equipped than was the case only a few years ago.

Nowadays the really serious amateur can equip himself with really sophisticated instrumentation, including CCDs (Charge-Coupled Devices), which extend the range beyond all recognition. I do not propose to go into details here, because I am not sufficiently expert; I admit to being an astronomical dinosaur of the pre-electronic age, and so I will confine myself to visual studies of the kind that I have been trying to carry out for the past sixty years or so.

For visual studies, I would say that the minimum useful size is an aperture of 4 inches (for a refractor) or 6 inches (for a Newtonian reflector), though of course smaller telescopes will give pleasing views, particularly of the Moon. My largest telescope at my Selsey observatory is a 15-inch reflector; it is big enough to be very useful indeed, though I appreciate that it has marked limitations.

Never try to use too high a magnification. A small, sharp image is always preferable to a larger, blurred one. A magnification of about ×50 to the inch of aperture is about as much as can normally be used even on good nights. I strongly disagree with the oft-expressed opinion that a small telescope can often show as much as a larger one, and in my experience the larger the telescope, the better the results (under good conditions, of course). I am equally opposed to stopping down an aperture. Be very careful about your choice of eyepieces, because using a good telescope with an inferior eyepiece is tantamount to using a good record player with a blunt needle.

Do not hurry an observation, and always record it as soon as it has been completed. The temptation to 'leave it until tomorrow' should be stubbornly resisted, as mistakes are bound to creep in.

Always write full notes. No observation is useful unless it carries the observer's name, time (GMT), telescope, magnification, and conditions of seeing. For this, the scale devised long ago by E. M. Antoniadi is to be recommended; it ranges from 1 (near-perfect conditions) down to 5 (very poor, so that no observation would be made except for some special purpose).

Never place any reliance upon an observation made under rushed or poor conditions. Such work is not only useless, but is actively misleading, as it will confuse subsequent analyses.

Above all, never give way to wishful thinking. Record only what has been seen with certainty, and beware of jumping to conclusions. If, for instance, you go to the telescope confident that you will see the north polar cap of Mars, you will probably 'see' it whether it is visible or not. Unconscious prejudice is difficult to avoid; the ability to do so is a supreme test of the observer's skill and experience.

Now let me turn to the planets one by one. I have given only very rough guidance; if you need something more detailed, turn to the observational notes given in books written specially for the purpose.

MERCURY

No useful work can be done with regard to the surface features of Mercury without the use of a telescope quite beyond the amateur's range. One effective programme, however, is to estimate the phase to see whether there is any discrepancy similar to the Schröter effect on Venus. The best method is to make a drawing at the telescope, and then measure it; the results are much better than a simple estimate. The main trouble is that observing Mercury is rather pointless except in daylight, or when the Sun has only just set or has not quite risen. To locate Mercury in daylight you need a good telescope equipped with accurate setting circles and clock drive. Never sweep aimlessly around searching for the planet; there is always the danger that the Sun will enter the field of view, with disastrous results. Binoculars are helpful in locating Mercury at dusk or dawn – but again, never sweep unless the Sun is completely below the horizon.

VENUS

Despite the general paucity of surface detail, Venus is always a worthwhile object to observe. The main amateur research programmes are:

1. Phase. I have described the Schröter effect in Chapter 6. As with Mercury, measure a drawing made at the telescope rather than be content with a simple estimate. Though the phase anomaly is most obvious near the time of dichotomy, it is generally present to some extent.
2. The shadings, which should be drawn as definitely as possible.
3. The bright areas, including the persistent cusp-caps.
4. Any sign of the Ashen Light. For this, it is essential to block out

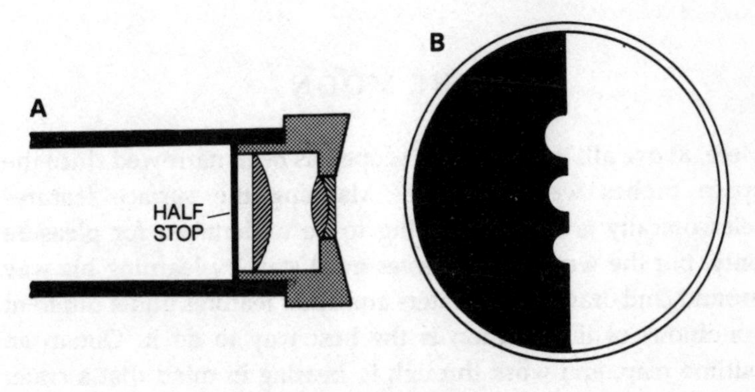

The Ashen Light eyepiece.

the bright crescent by some sort of occulting device in the eyepiece, because the observations have to be made against a darkish background; Venus is then inconveniently brilliant and also rather low down, so that seeing conditions are unlikely to be good.

For Ashen Light searches I use an eyepiece which has a built-in curved occulting section, as shown in the diagram. Fit the crescent of Venus into this curve, and only the dark side will be left – so that if there is any dim illumination you will see it. We are still not entirely confident about the origin of the Ashen Light, and there may be a connection with solar activity, as there certainly is with terrestrial auroræ. Obviously we are handicapped by the fact that searches are possible only during the crescent stage, and then only for a limited period each evening or morning when the sky is sufficiently dark and Venus is high enough above the horizon.

For other branches of investigation dusk and dawn conditions can be satisfactory, but many observers prefer to look at Venus in broad daylight. Luckily it is easy to find when well placed, and is visible in the average finder. If you have an equatorial mounting, a quick method is to set the declination of Venus and then swing *away* from the Sun by the difference in right ascension between the Sun and Venus.

Venus shows great variations in apparent diameter, but in general it is best to make all drawings to the same size. I use a diameter of 2 inches to the full circle.

Filters are an essential part of the equipment for observing Venus. Surface details are often enhanced, for instance, by the use of a yellow filter.

THE MOON

Here, above all, the amateur's scope has been narrowed since the space probes were launched. Mapping the surface features telescopically is now something to be undertaken for pleasure only, but the would-be observer must start by learning his way around, and drawing the craters and other features under different conditions of illumination is the best way to do it. Obtain an outline map, and work through it, bearing in mind that a crater which is prominent when shadow-filled at the terminator may be unidentifiable under high illumination, and that for most purposes full moon is the very worst time to start observing.

Certain relatively inconspicuous features, such as domes, can be usefully studied; the Orbiter and Apollo pictures cannot cover the Moon under all conditions of lighting, so that in some of the photographs low and gentle features do not show up. Then, of course, there are the TLP or Transient Lunar Phenomena, which take the form of short-lived glows or obscurations and indicate some sort of mild surface activity. Amateurs have played a major rôle in this research, and still do. A telescope of at least 6 inches aperture is needed, and it is useful to have a system of rotating filters known as a Moon-blink device. As always, the main hazard is unconscious prejudice. If you think that you have noted an unusual effect, check all adjacent areas to make sure that you have not been deceived by some quirk of the Earth's atmosphere, or instrumental trouble – and check the position and time very carefully. No TLP report is really valuable unless confirmed independently by another observer at a separate site, and in this work a few faulty reports can wreak havoc with all subsequent attempts at analysis.

MARS

Mars is a difficult object for a small aperture. It comes to opposition only in alternate years, and in general useful observations can be made for only a few months to either side of the opposition date. You really need a telescope of over 8 inches aperture to carry out much work with Mars, though of course the main features can be seen with a much smaller aperture.

Owing to the smallness of the disk, Mars is the one planet upon which a really high power should be used whenever possible. It is best to select a definite scale for the drawings (again I use a 2-inch circle for the full disk) and the phase, which is often considerable, should never be neglected. I always use prepared blanks; the actual phase can be looked up from yearly tables – and with Mars, there is no Schröter effect.

Before starting to draw Mars, it is wise to spend some time in simply looking at it, waiting until your eye has become fully adapted. When you begin drawing, sketch in the polar cap and the main dark areas as quickly as is compatible with accuracy, because Mars is rotating on its axis and there is a slow but perceptible drift of the markings across the disk. Minor details can then be added with the highest power which will give a sharp image. When the drawing is complete, and you are satisfied that nothing has been missed, add written notes about colours, intensities and any features of special interest, notably clouds.

Do not expect to see too much. At first you may be able to make out nothing apart from the polar cap (if it is present) and the main dark areas, but as you gain experience you will see more and more.

THE MINOR PLANETS

Only one asteroid (Vesta) can ever be seen with the naked eye, but quite a few are within range of binoculars, and dozens can be found with modest telescopes. Useful observations are of various kinds.

1. Magnitude estimates, using the same methods as those of variable star observers; check the asteroid against at least two comparison stars of known magnitude. There can be quite rapid variations, though usually they are so slight that some kind of photometer is needed.
2. Occultations of stars by minor planets. The duration of the occultation gives the diameter of the asteroid, and because the track of shadow is so narrow the amateur can be of tremendous help, because he can take his equipment to a suitable site. (It is seldom that an occultation track passes over a permanent observatory!) Watch out, too, for secondary occultations; there have been suspicions that some asteroids may be double.

JUPITER

It is no exaggeration to say that most of our really long-term knowledge of the changes on Jupiter's surface is due to amateur astronomers. The disk abounds in detail, and a modest telescope of aperture 6 inches or so is enough to enable the observer to do really valuable work.

Disk drawings should be made as quickly as possible, as the rate of spin is rapid and the drift of the markings is very obvious. A drawing should be completed within ten minutes at most with regard to the major details; the minor ones can then be filled in more slowly, without altering the general framework. The colours seen should be noted, and also the relative intensities of the belts and zones.

A very important part of the Jupiter programme is the taking of transits. A feature transits when it passes across the planet's central meridian, i.e. the line passing through both poles and the centre of the disk. The polar flattening enables the central meridian to be found easily, and this flattening should never be neglected when a drawing is to be made; I strongly recommend the use of prepared blanks.

Transits should be given to the nearest minute; accurate timing is essential. It might seem a difficult task to make estimates with sufficient precision, but with a little practice it becomes surprisingly easy, and it is possible to take many transits in the course of an hour or so when Jupiter is really active.

These transits are important because they allow the longitude of the feature to be calculated, and hence rotation periods to be derived. It is easy to work out the longitudes from the tables given in yearly publications such as the *Handbook* of the British Astronomical Association; nothing is involved apart from simple addition and subtraction.

As the equatorial zone (System I) rotates more quickly than the rest of the planet (System II), two sets of tables are needed, and there must be no confusion as to which set is used – otherwise the results will be most peculiar. (Remember, System I is bounded by the northern edge of the South Equatorial Belt and the southern edge of the North Equatorial Belt.) It is also necessary to indicate in which part of the belt the feature lies; for instance, NEBs indicates the southern part of the North Equatorial Belt. The usual abbreviations are: P = polar, T = temperate, Tr = tropical, Eq = equatorial, B = belt, Z = zone, D = disturbance, RS = Red Spot, pr = preceding, f = following. Here is a typical extract from my own observational notebook for one particular night:

1970 April 13/14. 12½in. reflector, × 360. Seeing 3 till 00.02; thereafter 3 to 4. NEB broad and dark, and STB also prominent, but the SEB was faint and double. The NTB was visible, and a darker section followed the transit at 00.28 (April 14). The Red Spot was much in evidence, and was highly coloured, but no sign of the Hollow. A bright narrow zone lay between the STB and the SSTB.

TRANSITS:	LONGITUDE I	LONGITUDE II
23.21 (Apr. 13) Pr. end of the Red Spot	...	017.9
23.25 F. end of projection from S. edge of NEB	023.9	...
23.34 Centre of Red Spot	...	025.7
23.27 White spot in N. of EqZ	031.2	...
23.41 Condensation in SSTB	...	029.9
23.46 F. end of Red Spot	...	032.9
00.02 (Apr. 14) Projection from S edge of STB	...	042.5
00.28 Pr. of dark section of NTB	...	058.2
02.03 Projection from S. edge of NEB	120.1	...
02.18 White spot in N. of EqZ	129.2	...

(Clouds, 23.48–00.00, 00.10–00.15, and 00.30–01.50.
Occultation of Io; 1st contact 23.54. Vanished, 23.56.
Emersion, 02.15 (definition poor at this time; seeing 4).
Clouds stopped observing for the night at 02.20.)

I do not claim that this was a good series of observations; I merely give it as a typical example of work on a rather poor night. I made a drawing of the disk at 23.37 to 23.47, and a strip drawing of the EqZ at 02.00 to 02.10.

So far as the Galilean satellites are concerned, it is useful to make careful timings of the transits (immersion and emersion at the planet's limb), shadow transits, occultations and eclipses.

SATURN

In some ways Saturn is a convenient planet. It bears high powers well, even better than Jupiter, and although there is not generally much surface detail there is always the chance of making a startling discovery such as a white spot. The paucity of well-defined detail means that surface transits are difficult to take, but whenever possible they should be observed in the same way as for Jupiter. For this, a telescope of at least 10 inches aperture is needed. It is also valuable to estimate the colours and intensities of the various zones on the planet, because erratic changes occur from time to time. For intensities, use a scale of from 0 (white) to 10 (black shadow).

Cassini's Division is easy when the rings are fairly well open, but Encke's Division is more elusive; look for it near the ansæ. Estimate the colours and intensities of the rings, using the same scale as for the disk. When a drawing is made, take care to put in the shadows accurately; globe upon rings, rings upon globe. Saturn is really at its most interesting, though least beautiful, when the rings are edgewise-on, as they will be in 1995.

Occasionally Saturn occults a star, and this is an important event, since it enables the transparencies of the various rings to be estimated.

Any 3-inch refractor will show Titan well, and with my 12½-inch reflector I have managed to see all the pre-Voyager satellites except Phœbe; obviously they are much easier with my 15-inch. The main interest is in estimating magnitudes. Convenient stars may be used for comparisons, but when there are no stars available, as often happens, the only real solution is to adopt a magnitude of 8.3 for Titan and work from that. Errors are bound to be introduced, but the method is better than nothing. Iapetus, of course, is of special interest.

URANUS

There should be no difficulty in identifying Uranus, and it is useful to make magnitude estimates. A low power is recommended, since with higher magnifications Uranus becomes so un-stellar in appearance that it cannot be compared with a star.

NEPTUNE

Neptune can be found with binoculars; its magnitude may be estimated in the same way as for Uranus, but I have to admit that little else can be done.

PLUTO

Pluto can be seen with a moderate telescope – anything above 8-in aperture will do – but it is difficult to identify, as it looks just like a faint star. Obtain magnitude estimates if you can.

I have said nothing here about amateur photography of the planets, which is admittedly rather difficult even though useful results can be obtained with Venus and Jupiter. The Moon, of course, is an ideal photographic subject, always provided that you are equipped with a good camera and a clock-driven telescope.

OCCULTATIONS

Occultation observations may not be as valuable now as they used to be, but they are still very useful – and here the amateur really comes into his own, because he is mobile.

Lunar occultations are fascinating to watch. Because the Moon has no atmosphere, the star shines steadily until the instant that it is covered, when it snaps out like a candle-flame in the wind. Accurate timings are still needed, and occasionally something unusual is found. During one occultation I saw, to my surprise, that the star 'faded out'. It proved to be a very close double star which had not been previously recognized as such.

Of special interest are grazing occultations, when the star skims past the lunar limb. Because the limb is so irregular, the star may vanish and reappear several times as it passes by mountains and valleys. This not only gives very accurate positioning, but helps in refining the contour of the lunar limb itself.

Occultations by planets are rare. Of the first magnitude stars only Aldebaran, Antares, Spica and Regulus are close enough to the ecliptic to be occulted; thus Regulus was occulted by Venus on 7 July 1959, and will again be occulted by Venus on 1 October 2044. Mutual planetary occultations are even rarer; the last was in 1818, when Venus occulted Jupiter, and the next occasion when this will happen will be on 22 November 2065. Venus, of course, has a thick and extensive atmosphere, so that an occulted star will flicker and fade for some seconds before disappearing.

In the text I have referred to the importance of occultations of stars by asteroids and by Pluto. These should always be observed – the trouble is that they happen so seldom!

APPENDIX II

Planetary Data

Planet	*Distance from the Sun, millions of miles* max	mean	min	*Orbital Period*	*Synodic Period, days*	*Axial Rotation (equatorial)*
Mercury	43	36	29	88 days	115.9	58.646 days
Venus	67	67.2	66.7	224.7 days	583.9	243.16 days
Earth	94.6	93	91.4	365.2 days	—	23h 56m
Mars	154.5	141.5	128.5	687 days	779.9	24h 37m
Jupiter	507	483	460	11.9 years	398.9	9h 50m
Saturn	938	886	835	29.5 years	378.1	10h 14m
Uranus	1867	1783	1699	84 years	369.7	17h 14m
Neptune	2817	2793	2769	164.8 years	367.5	16h 7m
Pluto	4583	3666	2766	247.7 years	366.7	6d 9h

Planet	*Orbital Eccentricity*	*Orbital Inclination, degrees*	*Axial Inclination, degrees*	*Diameter, miles (equatorial)*	*Apparent diameter, secs of arc* max	min
Mercury	0.206	7.0	2	3030	12.9	4.5
Venus	0.007	3.4	178	7523	66	9.6
Earth	0.017	0	23.4	7926	—	—
Mars	0.093	1.9	24	4222	25.7	3.5
Jupiter	0.048	1.3	3.1	89,424	50.1	30.4
Saturn	0.056	2.5	26.4	74,914	20.9	15
Uranus	0.047	0.8	98	31,770	3.7	3.1
Neptune	0.009	1.8	28.8	31,410	2.2	2
Pluto	0.248	17.1	122	1444	0.2	0.1

Planet	*Escape velocity, mi/sec*	*Mass, Earth = 1*	*Volume, Earth = 1*	*Density, water = 1*	*Maximum magnitude*
Mercury	2.6	0.055	0.056	5.44	−1.9
Venus	6.4	0.815	0.86	5.25	−4.4
Earth	7	1	1	5.52	—
Mars	3.2	0.107	0.15	3.94	−2.8
Jupiter	37	318	1319	1.3	−2.6
Saturn	22	95	744	0.7	−0.3
Uranus	14	14.6	67	1.3	+5.6
Neptune	15	17.2	57	2.1	+7.7
Pluto	0.7	0.002	0.007	2.02	+14

APPENDIX III

Satellite Data

Name	*Mean distance from centre of primary, thousands of miles*	*Orbital Period* *d*	*h*	*m*	*Orbital Eccentricity*
EARTH					
Moon	239	27	7	43	0.055
MARS					
Phobos	5.8	0	7	39	0.02
Deimos	14.6	1	6	18	0.003
JUPITER					
Metis	79.5	0	7	5	0
Adrastea	80.2	0	7	7	0
Amalthea	113	0	11	57	0.003
Thebe	138	0	16	12	0.013
Io	262	1	18	28	0.004
Europa	417	3	13	14	0.009
Ganymede	666	7	3	43	0.002
Callisto	1170	16	16	32	0.007
Leda	6895	239			0.148
Himalia	7135	251			0.158
Lysithea	7284	259			0.107
Elara	7295	260			0.207
Ananke	13,176	631			0.17
Carme	14,046	692			0.21
Pasiphaë	14,605	735			0.38
Sinope	14,730	758			0.28
SATURN					
Pan	82	0	13	41	0
Atlas	85.5	0	14	27	0.002
Prometheus	86.6	0	14	43	0.004
Pandora	88.1	0	15	6	0.004
Janus	94.1	0	16	41	0.007
Epimetheus	94.1	0	16	40	0.009
Mimas	115	0	22	37	0.020
Enceladus	148	1	8	53	0.004
Tethys	183	1	21	18	0
Telesto	183	1	21	18	0
Calypso	183	1	21	18	0
Dione	235	2	17	41	0.002

Name	Mean distance from centre of primary, thousands of miles	Orbital Period d	h	m	Orbital Eccentricity
Rhea	328	4	12	25	0.001
Titan	760	15	22	41	0.029
Hyperion	920	21	6	38	0.104
Iapetus	2200	79	7	56	0.028
Phœbe	8050	550	10	50	0.163
URANUS					
Cordelia	30.6	0	7	55	0
Ophelia	33.1	0	8	55	0
Bianca	36.7	0	10	23	0
Cressida	38.4	0	11	7	0
Desdemona	39	0	11	24	0
Juliet	40	0	11	50	0
Portia	41.1	0	12	19	0
Rosalind	43.5	0	13	24	0
Belinda	46.7	0	14	56	0
Puck	53.4	0	18	17	0
Miranda	81.1	1	19	50	0.017
Ariel	119	2	12	29	0.003
Umbriel	166	4	3	28	0.004
Titania	272	8	16	56	0.002
Oberon	365	13	11	7	0.001
NEPTUNE					
Naiad	29.8	0	7	6	0
Thalassa	31.1	0	7	30	0
Despina	32.6	0	8	0	0
Galatea	28.5	0	9	30	0
Larissa	45.7	0	13	18	0
Proteus	73.1	1	2	54	0
Triton	220.5	5	21	3	0.0002
Nereid	3460	359	21	7	0.749
PLUTO					
Charon	12.2	6	9	17	0

Name	Orbital inclination, degrees	Diameter, miles	Magnitude
EARTH			
Moon	5.1	2158	
MARS			
Phobos	1.1	17 × 14 × 11	11.6
Deimos	1.8	9 × 7 × 6	12.8

SATELLITE DATA

Name	*Orbital inclination, degrees*	*Diameter, miles*	*Magnitude*
JUPITER			
Metis	0	25	17.4
Adrastea	0	16 × 12 × 10	18.9
Amalthea	0.45	163 × 91 × 89	14.1
Thebe	0.9	68 × 56	15.5
Io	0.04	2264	5
Europa	0.47	1945	5.3
Ganymede	0.21	3274	4.6
Callisto	0.51	2981	5.6
Leda	26.1	6	20.2
Himalia	27.6	156	14.8
Lysithea	29	15	18.4
Elara	24.8	50	16.7
Ananke	147	12	18.9
Carme	164	19	18
Pasiphaë	145	22	17.7
Sinope	153	17	18.3
SATURN			
Pan	0	12	21
Atlas	0.3	23 × 21 × 17	18.1
Prometheus	0	92 × 62 × 42	16.5
Pandora	0.1	68 × 55 × 38	16.3
Janus	0.1	86 × 68 × 68	14.5
Epimetheus	0.3	120 × 118 × 96	15.5
Mimas	1.52	261 × 245 × 239	12.9
Enceladus	0.02	318 × 308 × 303	11.8
Tethys	1.86	650	10.3
Telesto	2	19 × 16 × 9	19
Calypso	2	19 × 10 × 10	18.5
Dione	0.02	696	10.4
Rhea	0.35	950	9.7
Titan	0.33	3201	8.3
Hyperion	0.43	224 × 174 × 140	14.2
Iapetus	7.52	892	10
Phœbe	175	143 × 137 × 130	16.5
URANUS			
Cordelia	0	16	
Ophelia	0	19	
Bianca	0	26	
Cressida	0	39	
Desdemona	0	34	below 21
Juliet	0	52	
Portia	0	67	
Rosalind	0	34	
Belinda	0	41	
Puck	0	96	

Name	*Orbital inclination, degrees*	*Diameter, miles*	*Magnitude*
Miranda	4.22	293	16.5
Ariel	0.031	720	14.4
Umbriel	0.036	727	15.3
Titania	0.014	981	14
Oberon	0.010	947	14.2
NEPTUNE			
Naiad	4.5	34	25
Thalassa	0	50	24
Despina	0	119	23
Galatea	0	92	23
Larissa	0	119	21
Proteus	0	258	20
Triton	159.9	1681	13.6
Nereid	27.2	149	18.7
PLUTO			
Charon	118	753	16.8

APPENDIX IV

Minor Planet Data

THE FIRST TEN ASTEROIDS

Asteroid	Mean distance from Sun, millions of miles	Orbital period, years	Diameter, miles	Rotation period, hours	Mean magnitude at opposition
1 Ceres	257	4.61	584	9.08	7.4
2 Pallas	257.4	4.62	360 × 292	7.81	8
3 Juno	247.8	4.36	179 × 143	7.21	8.7
4 Vesta	219.3	3.63	358	5.34	6.5
5 Astræa	239.3	4.14	75	16.81	9.8
6 Hebe	225.2	3.78	127	7.28	8.3
7 Iris	221.5	3.69	129	7.14	7.8
8 Flora	204.4	3.27	101	12.35	8.7
9 Metis	221.7	3.69	98	5.08	9.1
10 Hygeia	292.6	5.59	267	17.50	10.2

SOME OTHER INTERESTING ASTEROIDS

Asteroid	Mean distance from Sun, millions of miles	Orbital period, years	Diameter, miles	Rotation period, hours	Mean magnitude at opposition
132 Æthra	196.2	4.56	24	?	11.9
279 Thule	394.8	8.23	81	?	15.4
288 Glauke	229.7	4.58	19	1500	13.2
511 Davida	269.2	5.66	200	5.2	10.5
433 Eros	166.9	1.76	15	5.3	8.3 (max)
704 Interamnia	263.7	5.36	210	8.7	11
944 Hidalgo	365	14.15	17	10	15.3
951 Gaspra	186.9	3.28	15	20	14.1
1036 Ganymed	180.9	4.35	25	?	15
1566 Icarus	120.4	1.12	1	2.3	18
1221 Amor	140	2.66	11	?	18
1685 Toro	99.5	1.60	5	7.6	10.2
1862 Apollo	98.6	1.78	1	1.4	17
2060 Chiron	1255	50.68	150	?	17
2062 Aten	81.8	0.95	1	?	18
2100 Ra-Shalom	60.5	0.76	1	?	19
2340 Hathor	60.5	0.77	0.3	?	19
3200 Phæthon	65.6	1.43	3	4	17
5145 Pholus	1893	93	150	?	21

APPENDIX V

Astronomical Societies

Anyone with even a casual interest will do well to join an astronomical society. Lone observing is enjoyable, but working together with others is much more so; and for any useful research, co-operation is essential.

The leading amateur society in Britain is the British Astronomical Association (BBA), at Burlington House, Piccadilly, London W1. There are special sections devoted to the Moon; Mercury and Venus; Mars; Jupiter; Saturn, and Asteroids and Remote Planets, each controlled by an experienced Director. No qualifications other than interest and enthusiasm are needed for membership. Meetings are held regularly in London and at various provincial centres, and there are various publications, including a bi-monthly *Journal*, an annual *Handbook*, and irregular *Circulars* to give news of special events. The BAA has an observational record second to none.

There are many local societies of high standard, some with their own observatories; almost all large cities and many smaller towns and districts have them. A full list is given in the annual *Yearbook of Astronomy* (Sidgwick and Jackson).

APPENDIX VI

Bibliography

Many books have been written about the planets in recent years. It is clearly impossible to give a full list, so I have merely selected a few specialist books which are not too technical.

CATTERMOLE, P., *Planetary Volcanism*, Ellis Horwood, Chichester, 1989. *Mars*, Ellis Horwood, Chichester, 1992.

——, *Mars*, Chapman & Hall, London, 1993.

HOYT, W. G., *Planets X and Pluto*, University of Arizona Press, 1980.

HUNT, G., and MOORE, P., *The Planet Venus*, Faber and Faber, London, 1982. *Atlas of Uranus*, Cambridge University Press, 1989.

KOWAL, C. T., *Asteroids*, Ellis Horwood, Chichester, 1988.

MINER, E., *Uranus*, Ellis Horwood, Chichester, 1990.

MOORE, PATRICK, *The Planet Neptune*, Ellis Horwood, Chichester, 1988.

PEEK, B., *The Planet Jupiter*, Faber and Faber, London, 1981.

RÜKL, A., *Atlas of the Moon*, Hamlyn, London, 1990.

STROM, R. G., *Mercury*, Cambridge University Press, 1987.

TOMBAUGH, CLYDE, and MOORE, PATRICK, *Out of the Darkness: the Planet Pluto*, Lutterworth Press, London, and Stackpole Books, New York, 1980.

various. *Guide for Observers of the Moon*, published by the British Astronomical Association, London 1989.

All these books are in print. Unfortunately there are some very valuable books which are out of stock – notably *The Amateur Astronomer's Lunar Atlas*, by H. R. Hatfield (Lutterworth, 1968), which is still in circulation in libraries and is ideally suited to the amateur observer. This should certainly be reprinted.

The main monthly periodicals are *Astronomy Now* (in Britain), published by Intra Press in London, and *Sky and Telescope* in the United States. These can be obtained from any good newsagent. Many local societies publish journals of their own.

INDEX